Sustainable Aviation - Multidisciplinary Approaches, Technologies, and Strategies

Edited by Melih Cemal Kuşhan

Published in London, United Kingdom

Sustainable Aviation - Multidisciplinary Approaches, Technologies, and Strategies
http://dx.doi.org/10.5772/intechopen.1009699
Edited by Melih Cemal Kuşhan

Contributors
Abderrahim Nabi, Abdul Hai, Caroline Willich, Christiane Bauer, Daniel Frank, Djamal Darfilal, Fatima AlHammadi, Hamed Alhashmi, Judith de Patoul, Melih Cemal Kuşhan, Roland Ademi, Salma S. Syed, Samara Brandão de Almeida, Sean Shan Min Swei, Tobias G.W. Verhulst, Tobias Graf, Woojin Choi, Yana Maneva

First published in London, United Kingdom, 2026 by IntechOpen
IntechOpen is the global imprint of INTECHOPEN LIMITED, registered in England and Wales, registration number: 11086078, 167-169 Great Portland Street, London, W1W 5PF, United Kingdom

For EU product safety concerns: IN TECH d.o.o., Prolaz Marije Krucifikse Kozulić 3, 51000 Rijeka, Croatia, info@intechopen.com or visit our website at intechopen.com.

British Library Cataloguing-in-Publication Data
A catalogue record for this book is available from the British Library

Sustainable Aviation - Multidisciplinary Approaches, Technologies, and Strategies
Edited by Melih Cemal Kuşhan
p. cm.
Print ISBN 978-1-83635-992-0
Online ISBN 978-1-83635-991-3
eBook (PDF) ISBN 978-1-83635-993-7

Meet the editor

Prof. Dr. Melih Cemal Kushan graduated from the Department of Mechanical Engineering at Dokuz Eylül University, Turkiye, in 1985. He received his MSc degree in Industrial Engineering from Anadolu University in 1988 and his Ph.D. degree in Mechanical Engineering from Eskisehir Osmangazi University (ESOGÜ) in 1996. In 2009, he completed postdoctoral research in Mechanical Engineering at James Cook University, Australia. He is currently a Professor in the Department of Aircraft Engineering at Eskisehir Osmangazi University. He also serves as the Head of the Department of Aircraft Engineering and Director of the Aviation and Space Technologies Application and Research Center (HUZTAM). Prof. Dr. Kushan has authored and co-authored 98 journal articles, 137 conference papers, and 39 books and book chapters. He has participated in 24 research and development projects funded by organizations including the European Union, TÜBİTAK, and KOSGEB. His publications have received more than 3,600 citations and cover a wide range of topics in aerospace engineering, materials science, aircraft structures, aviation technologies, and sustainable aviation.

Contents

Preface

Aviation has transformed the way humanity connects, travels, conducts business, and exchanges knowledge. Over the past century, remarkable advances in aerodynamics, materials, propulsion systems, manufacturing technologies, and operational strategies have enabled unprecedented levels of mobility and global accessibility. However, alongside these achievements, the aviation sector faces growing challenges related to environmental sustainability, energy consumption, resource utilization, and climate impact. Addressing these challenges requires a multidisciplinary perspective that integrates engineering innovation, operational efficiency, environmental responsibility, and strategic planning.

The edited volume *Sustainable Aviation – Multidisciplinary Approaches, Technologies, and Strategies* has been prepared to contribute to the ongoing global dialogue on the future of sustainable air transportation. The objective of this book is to present a comprehensive overview of contemporary developments, emerging technologies, and innovative approaches that support the transition toward a more sustainable aviation industry.

Sustainability in aviation should not be viewed merely as an environmental obligation but as a fundamental design driver influencing every stage of an aircraft's lifecycle. From conceptual design and material selection to manufacturing, operation, maintenance, and end-of-life management, sustainability considerations increasingly shape engineering decisions and strategic policies. Consequently, achieving meaningful progress requires collaboration among researchers, engineers, policymakers, industry professionals, and academic institutions.

The chapters included in this volume address a wide spectrum of topics related to sustainable aviation. They explore advancements in aerospace materials, lightweight structures, alternative propulsion concepts, energy-efficient technologies, manufacturing processes, maintenance strategies, operational improvements, environmental assessment methods, and emerging technological solutions. Together, these contributions illustrate how multidisciplinary research can support the aviation sector in reducing its environmental footprint while maintaining safety, reliability, and economic viability.

The diversity of perspectives represented in this book reflects the inherently interdisciplinary nature of sustainable aviation. The contributing authors bring expertise from various fields and geographical regions, providing valuable insights into both current challenges and future opportunities. Their collective efforts demonstrate that sustainability is not a single technological solution but rather a comprehensive framework requiring continuous innovation and collaboration.

As the editor of this volume, I hope that this book will serve as a useful resource for researchers, graduate students, industry professionals, policymakers, and all readers interested in the future of aviation. The discussions presented herein highlight the importance of integrating technological advancement with environmental stewardship and responsible decision-making.

I would like to express my sincere gratitude to all contributing authors for sharing their expertise and research perspectives. I also extend my appreciation to the reviewers and editorial staff whose efforts have contributed significantly to the development and completion of this volume. Their dedication and professionalism have been essential in bringing this project to fruition.

It is my hope that the ideas and discussions presented in this book will inspire further research, foster interdisciplinary collaboration, and support the development of innovative solutions for a more sustainable aviation future.

Melih Cemal Kushan
Department of Aircraft Engineering,
Eskisehir Osmangazi University,
Eskisehir, Turkiye

Chapter 1

Introductory Chapter: Sustainable Aviation – Materials, Technologies and System-Level Transformation

Melih Cemal Kuşhan

1. Introduction

Aviation represents one of the most influential engineering domains, historically shaped by performance-driven design priorities. For decades, efficiency, safety, and operational capability have guided technological advancements, often outweighing environmental considerations. Today, however, this paradigm is undergoing a fundamental transformation.

By enabling rapid global mobility, economic integration, and cultural exchange, aviation has become an indispensable component of modern society. At the same time, its environmental footprint, arising from both CO2 and non-CO2 emissions such as nitrogen oxides and contrail formation, has positioned sustainability as a defining challenge for the future of aeroplane systems [1].

In response to increasing regulatory pressure, societal expectations, and global net-zero commitments, the aviation sector is transitioning beyond incremental improvements toward a more comprehensive and multidisciplinary transformation. This shift encompasses not only technological innovation but also the integration of materials science, manufacturing processes, system architectures, and lifecycle management strategies.

From an editorial standpoint, sustainability in aviation should be interpreted not merely as a constraint, but as a primary design driver that fundamentally reshapes engineering priorities across materials, manufacturing, and system-level integration.

2. The sustainability challenge in aviation

The sustainability challenge in aviation is inherently multi-dimensional, requiring a balance between environmental responsibility, safety, economic viability, and operational performance. Unlike many other transportation sectors, aviation systems operate under stringent certification requirements and extreme design constraints, where even minor weight increases can have significant operational consequences.

In addition to carbon emissions, non-CO2 effects, particularly contrail formation and nitrogen oxide emissions, play a critical role in atmospheric warming. These combined effects highlight the necessity of adopting a systems-level approach, where sustainability is embedded into the design process rather than treated as an external constraint.

Furthermore, the long operational lifespan of aeroplane platforms introduces additional complexity. Sustainable transformation must therefore address both the development of next-generation technologies and the optimization of existing fleets through maintenance, repair, and operational strategies.

3. Materials and manufacturing transformation

The evolution of materials and manufacturing technologies represents one of the most significant drivers of sustainability in aviation. The transition from conventional metallic structures to advanced composite systems has enabled substantial reductions in structural weight, directly contributing to improved fuel efficiency and reduced emissions.

Recent developments in multifunctional composite materials further extend these benefits. For example, eco-friendly cork-based composites offer enhanced vibration damping and thermal insulation while reducing environmental impact [2]. Similarly, composite systems enhanced with shear thickening fluids provide improved impact resistance and energy absorption, demonstrating how material systems can simultaneously address performance, safety, and sustainability requirements [3, 4].

In parallel, additive manufacturing technologies have evolved from prototyping tools into key enablers of structural optimization in aerospace applications. Their ability to produce complex geometries with reduced material waste makes them particularly suitable for lightweight design strategies, which are essential for minimizing fuel consumption and environmental impact [5, 6].

As illustrated in **Figure 1**, sustainable aviation requires integrated interactions between materials, manufacturing technologies, system-level engineering, and lifecycle management approaches.

Advanced post-processing techniques, such as hot isostatic pressing, further enhance material integrity and mechanical performance, extending component life and contributing to lifecycle sustainability [7].

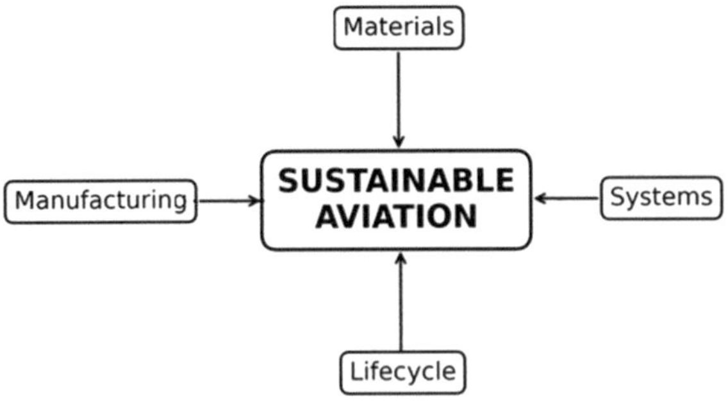

Figure 1.
Conceptual framework illustrating the interaction between materials, manufacturing, system-level design, and lifecycle considerations in sustainable aviation?.

4. System-level transformation in aviation

While material and manufacturing innovations are critical, sustainable aviation cannot be achieved without a broader system-level transformation. This includes advancements in propulsion technologies, aerodynamic design, and digital engineering methodologies.

Electric and hybrid-electric propulsion systems are emerging as promising solutions for reducing emissions, particularly in short – and medium-range applications [8, 9]. At the same time, improvements in aerodynamic efficiency, enabled by advanced computational tools, continue to reduce energy consumption during flight operations.

Digitalization plays a central role in this transformation. Technologies such as digital twins, data-driven maintenance strategies, and advanced simulation frameworks enable more efficient design, operation, and lifecycle management of aeroplane systems [10].

5. Lifecycle perspective and future outlook

Sustainability in aviation must ultimately be evaluated from a lifecycle perspective, encompassing material extraction, manufacturing, operation, maintenance, and end-of-life processes. This holistic approach requires interdisciplinary collaboration across materials science, structural engineering, manufacturing technologies, and environmental assessment [11].

Looking forward, sustainable aviation will likely be shaped by the convergence of multiple technological domains rather than a single breakthrough innovation. Lightweight and multifunctional materials, advanced manufacturing techniques, cleaner propulsion systems, and intelligent operational strategies will collectively define the next generation of aeroplane systems.

From an editorial perspective, sustainable aviation can only be effectively addressed through a multidisciplinary framework that integrates materials innovation, advanced manufacturing, system-level design, and lifecycle considerations. Such an approach highlights the necessity of moving beyond isolated technological solutions toward a more holistic engineering paradigm capable of addressing the complex environmental challenges of modern aviation.

AI tools disclosure

Artificial intelligence (AI)-assisted tools were used only for limited language editing, grammar improvement, and formatting support during the preparation of this chapter. All scientific interpretations, technical evaluations, conceptual development, and editorial decisions were conducted solely by the author. The author takes full responsibility for the accuracy, originality, and integrity of the content presented in this work.

Author details

Melih Cemal Kuşhan
Eskişehir Osmangazi University, Eskişehir, Turkey

*Address all correspondence to: erzesk@gmail.com

References

[1] Lee DS, Fahey DW, Skowron A, et al. The contribution of global aviation to anthropogenic climate forcing. Atmospheric Environment. 2021;**244**:117834. DOI: 10.1016/j.atmosenv.2020.117834

[2] Gurgen S, Fernandes FAO, Sousa RJA, Pereira A, Verdum M, Mares A, Ptak M, Sawicki M, Leonaviciute V, Nugaras J, Kushan MC, Sofuoglu A, Gurgen S. Sustainable and eco-friendly cork composites in aerospace engineering. In: Proceedings of TEchMA2022 – 5th International Conference on Technologies for Wellbeing and Sustainable Manufacturing Solutions; Portugal: University of Aveiro; 2022

[3] Gurgen S, Kushan MC. Improvement of spall liner performance with smart fluid applications. Thin-Walled Structures. 2022;**180**:109854. DOI: 10.1016/j.tws.2022.109854

[4] Sheikhi MR, Gurgen S, Kushan MC. Design of smart sandwich structures enhanced by multifunctional shear thickening fluids. Composite Structures. 2023. DOI: 10.1016/j.compstruct.2023.117520

[5] DebRoy T, Wei HL, Zuback JS, Mukherjee T, Elmer JW, Milewski JO, Beese AM, Wilson-Heid A, De A, Zhang W. Additive manufacturing of metallic components-Process, structure and properties. Progress in Materials Science. 2018;**92**:112–224. DOI: 10.1016/j.pmatsci.2017.10.001

[6] Karasoglu M, Öteyaka MÖ, Yasa E, Tan E, Kuşhan MC. Effect of heat treatment and hot isostatic pressing on the corrosion behavior of Ti6Al4 V parts produced by electron beam melting additive manufacturing technology. ACS Omega. 2024;**9**(27):29904–29916. DOI: 10.1021/acsomega.4c04218

[7] Karasoglu M, Yasa E, Tan E, Kavak S, Soylemez E, Kushan MC. Influence of heat treatments and hot isostatic processes (Hip) on microstructure and mechanical behavior of Ti-6Al-4V alloy produced by Powder Bed Fusion Using Electron Beam (PBF-EB). Kovove Materialy. 2025;**63**(5-6):331–356. DOI: 10.31577/km.2025.5-6.331

[8] Gudmundsson S. General Aviation Aircraft Design: Applied Methods and Procedures. Oxford, UK: Butterworth-Heinemann; 2014

[9] Jansen RH, Bowman C, Jankovsky A, Dyson R, Felder J. "Overview of NASA Electrified Aircraft Propulsion (EAP) Research for Large Subsonic Transports." In: AIAA 2017-4701. 53rd AIAA/SAE/ASEE Joint Propulsion Conference; July 2017 Georgia, USA: Atlanta. DOI: 10.2514/6.2017-4701

[10] Shafto M, Conroy M, Doyle R, Glaessgen E, Kemp C, LeMoigne J, Wang L. Modeling, Simulation. Washington, USA: Information Technology & Processing Roadmap. National Aeronautics and Space Administration (NASA); 2012

[11] Ashby MF. Materials and the Environment: Eco-informed Material Choice. 2nd ed. Oxford, UK: Butterworth-Heinemann; 2013

Chapter 2

Propelling Sustainability: Green and Alternative Fuels for Space Missions

Abdul Hai, Djamal Darfilal, Fatima AlHammadi, Salma S. Syed, Hamed Alhashmi, Abderrahim Nabi and Sean Shan Min Swei

Abstract

The quest for green space exploration has accelerated interest in green propulsion and alternative fuel systems. Conventional chemical propellants, such as hydrazine, have high performance but cause adverse environmental and safety impacts. Less toxic and more environmentally friendly are the green propellants (e.g., hydroxylammonium nitrate (HAN), ammonium dinitramide (ADN), hydrogen peroxide (high-test peroxide [HTP]), and liquid oxygen–liquid methane (LOX–CH_4)) for sufficient propulsion capability with relatively safe handling. This chapter presents the principal physicochemical properties, performance metrics (e.g., specific impulse and density impulse), and material compatibility of these emerging propellants. Their use in monopropellant, bipropellant, and hybrid propulsion systems is assessed for next-generation spacecraft design. In the context of long-stay and return missions, the possibility of *in situ* resource utilization (ISRU), specifically the production of propellants from lunar and Martian materials, is presented. Finally, electric and hybrid propulsion systems are discussed as potential complementary options for low-thrust, long-time-of-flight missions. Key challenges, including material degradation, complexity and scalability of storage, and technology readiness, are presented with future research directions being outlined. By combining environmental and regulatory dimensions, the chapter presents a comprehensive summary of the factors and considerations that could affect green propulsion technologies.

Keywords: green propellants, alternative fuels, sustainable space mission, electric and hybrid propulsion, *in situ* resource utilization (ISRU), environmental impact

1. Introduction

The rapid expansion of space activities has brought increased attention to the environmental effects of propulsion systems, as these systems have historically emphasized performance over sustainability [1]. Green and alternative propellants are becoming more popular solutions for meeting efficiency requirements and ecological responsibility standards due to rising climate concerns, as well as evolving industry expectations and regulations. An important aspect of green propellants (e.g., high-test peroxide "HTP") is their safer handling compared to conventional propellants, such as hydrazine [1]. The main components of conventional space propulsion systems consist of hypergolic propellants, which include hydrazine, along with monomethylhydrazine (MMH) and nitrogen tetroxide (NTO) [2]. These substances provide high energy density and storability but are toxic and carcinogenic while persisting in the environment. The use of traditional chemical propellants demands elaborate handling procedures and specific safety protocols, while producing exhaust gases that pollute the atmosphere and deplete the ozone layer; these problems worsen with increasing launch rates [3]. An additional advantage of green propellants is that they make propulsion technology accessible to academic institutions and emerging countries, whereas conventional systems require heavy investment and a strong industrial base. The expansion of reusable launch systems, together with satellite constellations and deep-space missions, requires environmental impact reduction to match performance and reliability standards [4]. The Environmental, Social, and Governance (ESG) frameworks now shape aerospace operations by requiring public and private sectors to incorporate sustainability principles into their mission planning and execution. The European Space Agency (ESA), along with the National Aeronautics and Space Administration (NASA) and other stakeholders, now includes environmental assessments as a mandatory component of their propulsion technology roadmaps [5, 6]. The development of new propulsion alternatives is essential, as environmental regulations now affect manufacturing operations, fuel handling, and atmospheric emissions.

The growing environmental awareness has led to increased adoption of green propulsion technologies for sustainable space exploration [7]. The range of alternative propulsion systems includes HTP and liquid oxygen (LOX)-based systems, together with advanced hybrid propulsion models that use less hazardous oxidizers and fuels [8, 9]. The propellants present dual advantages of environmental protection and enhanced safety during handling, as well as lower lifecycle expenses, while enabling *in situ* resource utilization (ISRU) strategies, which prove essential for extended lunar and Martian missions. The LOX/methane (LOX–CH_4) propulsion system provides environmentally friendly combustion while allowing for the extraction of fuel components from extraterrestrial resources, which supports closed-loop life support and logistics in off-world habitats [10]. Green and alternative propulsion systems face essential technical and logistical challenges that need resolution before being deployed on a large scale.

This chapter presents green and alternative propulsion technologies for sustainable space exploration. It discusses traditional chemical propulsion systems alongside emerging green alternatives through an analysis of scientific principles and the benefits and limitations of different propellant types. The chapter investigates the role of green propellants within monopropellant, bipropellant, and hybrid propulsion systems to determine their applicability for future spacecraft designs. It also examines the role of ISRU, as well as electric and hybrid propulsion working together to support deep-space return missions and long-duration missions. The evaluation of green propellants through technical performance analysis, environmental impact assessment, and regulatory framework examination establishes a multidisciplinary base for understanding their application in future mission planning. The research aims to educate engineers, policymakers, and researchers about the trade-offs and opportunities in adopting cleaner propulsion systems to develop sustainable space operations.

2. Overview of conventional propulsion systems

Spaceflight operations rely on chemical propulsion, enabling launch and orbital transfer, as well as deep-space operations using solid, liquid, and hybrid propellant systems. Solid propellants serve multiple applications in boosters and missiles because of their simple systems and high thrust levels. After ignition, solid propellants exhibit limited control capabilities, while their energetic mixtures create environmental and safety issues [11]. Liquid bipropellant systems using hydrazine with NTO provide better specific impulse and improved thrust control, making them perfect for upper-stage propulsion and in-space maneuvering applications. The operational benefits of liquid systems come with requirements for extensive ground infrastructure and dangerous handling practices due to their toxic nature [12]. The integration of liquid oxidizers with solid fuel in hybrid propulsion systems shows theoretical advantages for throttle control and safety, yet their development and flight experience remain constrained. The chemical propellant family of hydrazine and its derivatives, MMH and unsymmetrical dimethylhydrazine (UDMH), has gained widespread adoption because of its storage capabilities and reliable ignition when working with existing spacecraft equipment [13]. These substances have served as power sources for satellite attitude control systems, orbital adjustments, and planetary probes throughout history. The hazardous chemical properties of these substances require extensive safety precautions due to their toxic, carcinogenic nature and instability. Hydrazine-based systems need complete personal protective equipment for full-body protection, as well as specialized fueling equipment and remote handling facilities for safety reasons. The combustion process of these substances generates hazardous byproducts, including nitrogen oxides and volatile organic compounds (VOCs), leading to atmospheric deterioration and possible long-term ecological damage [14]. Traditional propellant usage results in growing health and safety risks and environmental problems, which have led to tighter regulatory requirements. The United Nations Recommendations on the Transport of Dangerous Goods, the Occupational Safety and Health Administration (OSHA; USA), and the European Union Registration, Evaluation, Authorisation and Restriction of Chemicals (EU REACH) framework classify hydrazine, MMH, UDMH, and NTO as hazardous materials under international and national regulations [15]. The regulatory requirements establish detailed procedures for storage, handling, transportation, and disposal operations. It is worth noting that mission operators experience increasing operational expenses together with administrative complexities due to performing detailed safety evaluations and environmental impact assessments, employing certified personnel, and creating contingency plans. The advancing regulatory environment drives the space industry to adopt "green" propellants as its primary choice. New propellant technologies aim to match traditional performance characteristics while reducing toxicity levels and simplifying handling procedures, all while fulfilling worldwide sustainability standards. NASA and the ESA conduct official programs to eliminate hydrazine through research funding and flight testing of environmentally friendly propellants [16]. The existing mission architectures rely on legacy propulsion systems, yet their prolonged use becomes increasingly uncertain because of growing environmental risks, rising regulatory pressures, and increasing operational expenses. Space missions of the next generation require green propulsion technologies because regulatory compliance, operational efficiency, and environmental stewardship have become essential strategic priorities. It is worth highlighting the latest UNOOSA assembly report, which, in 2025, stressed the importance of promoting environmentally sustainable green propulsion technologies.

3. Green propellants: Definitions and criteria

3.1 What qualifies as a green propellant?

The green propellants include chemical substances and formulations that serve spacecraft propulsion while protecting the environment through safe operations and non-toxic properties without losing performance capabilities. Green propellants differ from traditional hydrazine propellants due to lower toxicity levels to humans and ecosystems, simpler handling procedures, and reduced regulatory requirements [17]. A green propellant must fulfill four essential criteria, which include non-carcinogenic and non-corrosive properties, together with low vapor pressure for inhalation safety and minimal hazardous exhaust emissions [18]. Green propellants need to maintain their properties over extended storage periods under standard conditions while being compatible with standard or slightly modified propulsion equipment for affordable implementation. The sustainability goals of green propellants are achieved through renewable feedstocks and ISRU compatibility for long-duration missions. The main green propellant examples include hydroxylammonium nitrate (HAN)-based monopropellants (e.g., AF-M315E), ammonium dinitramide (ADN)-based formulations (e.g., LMP-103S), and hydrogen peroxide (HTP) at high concentrations [19]. These alternatives provide multiple benefits, including safer handling procedures, reduced launch-site infrastructure requirements, and better compliance with modern environmental and occupational standards [19]. The evaluation process for green propellant status requires equal consideration of environmental aspects alongside operational and technical factors, while sustainability frameworks in aerospace engineering are becoming increasingly important for full life-cycle assessment.

3.2 Performance metrics: *Isp*, density impulse, storability

The performance of the propellants was assessed in terms of specific impulse (*Isp*), density impulse, and storability [20]. These terms are discussed in detail as follows:

3.2.1 Specific impulse (Isp)

Specific impulse (denoted as *Isp*) is the primary metric for evaluating the efficiency of a propellant in generating thrust. It represents the impulse (thrust × time) delivered per unit mass of propellant and is typically expressed in seconds. A higher *Isp* indicates a more efficient propulsion system.

$$Isp = \frac{F}{\dot{m} * g_o}$$

where F is the thrust (N), $\dot{m}$ represents the propellant mass flow rate (kg/s), and g_o is the standard gravity (9.80665 m/s^2).

Conventional propellants like hydrazine offer *Isp* values around 220–235 seconds in monopropellant systems, while sustainable alternatives, such as MMH/N_2O_4 and new rocket green propellant (NRGP), typically achieve *Isp* values of 341 and 328 seconds, respectively, demonstrating comparable or superior efficiency [14].

3.2.2 Density impulse

The volumetric performance of a propellant can be determined by the density impulse, which combines specific impulse (*Isp*: seconds) with density (ρ: kg/m^3). This metric is especially important for space missions where tank volume is a critical constraint. Some green propellants have higher densities than hydrazine; for example, AF-M315E exhibits a density of around 1.47 g/cm^3 compared to hydrazine's 1.01 g/cm^3 [21], resulting in a higher density impulse and thus more compact propulsion system designs with increased total impulse for a given tank volume.

3.2.3 Storability

The storability of a propellant describes its capability to preserve chemical stability along with performance characteristics when stored in space or on Earth for long durations. Some green propellants were previously discarded because of their unstable storage characteristics. With recent advances in materials, especially the emergence of highly compatible composites (e.g., HTP), interest in these propellants has been revitalized. The factors that affect storability include boiling point, vapor pressure, and chemical reactivity. The storability evaluation of green propellants requires assessment of thermal stability, material compatibility, low vapor pressure, and operational stability across different mission durations and thermal conditions [15]. The green propellants LMP-103S and AF-M315E demonstrate extended storage capabilities and stable thermal and chemical properties, which make them suitable for long-term space missions with simplified ground operations. The evaluation of green propellants through these performance metrics provides a complete understanding of their operational readiness [19]. The advantages of density impulse and storability in certain alternatives surpass their slightly lower *Isp* performance, especially when safety requirements, compactness, and regulatory compliance are essential mission factors.

3.3 Environmental and operational benefits

Green propellants offer multiple advantages, solving various problems associated with traditional chemical propulsion systems in terms of environmental performance and operational capabilities. From an environmental perspective, one of the most significant benefits is the reduction in toxicity and ecological risk. The environmental risk and ecological toxicity of traditional hydrazine derivatives stand in stark contrast to AF-M315E, LMP-103S, and high-concentration hydrogen peroxide (HTP), which demonstrate low environmental stability while producing cleaner combustion waste [22]. The low environmental persistence rates, combined with the safe combustion of these propellants, create both lower risks of soil and groundwater contamination and reduced dangerous nitrogen oxide emissions. Some green propellants can support ISRU operations by working with local or renewable oxidizers and fuels in extraterrestrial settings to establish closed-loop life support systems and extend mission duration [23]. The operational advantages of these systems match their environmental advantages. Some of the green propellants possess greater density levels than hydrazine, which leads to better volumetric efficiency and increased volume-specific impulse for more compact propulsion system designs. The operational benefits from these alternatives include superior handling properties because they exhibit lower vapor pressure, better thermal stability, and reduced corrosiveness, which simplifies ground support operations and improves safety during fueling and maintenance. The use of green propellants enables missions to benefit from enhanced logistics simplicity while also achieving shorter launch preparation periods and enhanced crew protection, which becomes crucial for crewed space missions and missions near sensitive payloads [24]. The reduced hazard classification of green propellants leads to easier regulatory compliance since they require fewer safety precautions during storage and transportation, thus simplifying REACH compliance in Europe and OSHA compliance in the United States [25, 26]. These propellants support the ESG criteria that aerospace industry operators increasingly adopt because of their ability to reduce both environmental harm and toxic risks. NASA, together with the ESA, has executed successful spaceflight tests of green propulsion systems, which proved both their operational feasibility and their environmental benefits in practical space missions [4]. The implementation of green propellants in mission architecture leads to improved space exploration through more sustainable and cost-effective operations. The increasing need for reusable launch systems, alongside responsive space access and lunar and Mars exploration, requires green propulsion systems, providing effective solutions that balance performance requirements and environmental protection for Earth and space.

4. Key green propulsion technologies

4.1 HAN systems

HAN was first investigated in the mid-twentieth century because of its potential as an energetic oxidizer and monopropellant ingredient, offering safer chlorine-free alternatives to hydrazine-based propellants. Research on HAN for military and aerospace propulsion continued at US and European research facilities because it dissolves well and forms ionic liquid blends with fuels. HAN-based propellants gained popularity as green propellants during the last several decades due to their environmental advantages and reduced toxicity. It represents a leading class of green monopropellants, replacing toxic legacy substances such as hydrazine. HAN is a water-soluble, ionic compound that acts as an oxidizer in energetic formulations and is often blended with fuel components, stabilizers, and combustion enhancers [8]. The most widely studied HAN-based formulation is AF-M315E, which was developed under NASA's Green Propellant Infusion Mission (GPIM) [27]. This propellant offers several operational, safety, and performance advantages over hydrazine, making it a viable alternative for satellite propulsion, orbital maneuvering, and spacecraft attitude control. In terms of performance, HAN-based propellants typically achieve a higher specific impulse (*Isp*) of 257 seconds and density impulse (~1.47 g/cm^3) than hydrazine, that is, 235 seconds and ~1.01 g/cm^3, respectively [27]. This results in a higher total impulse per unit volume, enabling more compact tank designs and increased mission payload capacity. From a safety perspective, HAN systems are substantially less toxic than hydrazine, reducing health hazards for ground personnel and eliminating the need for extensive protective equipment and complex fueling procedures. HAN has a much higher decomposition temperature and is less volatile, which enhances storage stability and reduces unintentional ignition risks. These characteristics simplify handling logistics and can lower overall mission costs through reduced infrastructure and compliance burdens. Operationally, HAN-based systems support longer mission durations and are compatible with electrically heated catalytic thrusters, offering precise thrust modulation and reliable on-orbit performance. The successful demonstration of AF-M315E on NASA's GPIM mission in 2019 confirmed its readiness for flight and opened the path for broader adoption across government and commercial platforms [73]. Overall, HAN-based propulsion systems present a balanced solution that improves performance while addressing the critical environmental, safety, and regulatory challenges posed by traditional chemical propellants. As demand for green, high-efficiency alternatives grows, HAN formulations are expected to play a central role in the next generation of spacecraft propulsion.

4.2 ADN formulations

ADN was discovered in the late 1970s by Soviet researchers and later developed in Sweden during the 1990s as a high-energy, chlorine-free oxidizer for propellants. ADN-based propellants constitute a crucial development in green propulsion technology as an environmentally friendly substitute for traditional hydrazine-based systems. The high-energy oxidizer ADN creates a liquid monopropellant through its combination with appropriate fuel components and stabilizers, demonstrating favorable performance, safety features, and handling characteristics [28]. The Swedish Space Corporation, together with ESA, developed LMP-103S, which represents the most recognized ADN-based formulation that successfully flew on Prisma and SkySat missions [29]. Performance data show that ADN-based propellants generate *Isp* between 230 and 250 seconds, which exceeds or matches the performance of hydrazine. The high density of LMP-103S at 1.24 g/cm^3 enables dense impulse production, resulting in more compact and efficient propulsion system designs [29]. The benefits of compactness and efficiency directly improve satellite platform performance because small satellite platforms have limited space availability. The safety profile and environmental impact of ADN-based propellants exceed those of hydrazine because they possess lower toxic and volatile characteristics. The propellant shows minimal vapor emission; is non-carcinogenic; and poses minimal risks during storage, transportation, and handling operations [30]. The absence of protective equipment, ventilation systems, and hazardous materials protocols in ground operations using ADN systems leads to faster turnaround times and reduced operational costs. These advantages align with current environmental regulations and aerospace safety standards. The operation of ADN formulations in monopropellant thrusters requires minimal modifications because they work well with existing propulsion architectures. The propellant maintains reliable long-term stability and performance consistency during space missions that operate in Earth orbit and beyond [31]. The elevated combustion temperature, together with superior thermal characteristics, allows for better energy conversion and thrust management. The combination of competitive performance with environmental compliance, operational safety, and system integration flexibility makes ADN-based propellants, including LMP-103S, an optimal green replacement for traditional fuels [32]. ADN formulations have become essential components for future propulsion systems because sustainable spaceflight has become a strategic focus of the industry.

4.3 Liquid oxygen–methane

LOX–CH_4 propulsion has been studied since the 1960s but only gained significant momentum in the early 2000s with the rise of reusable launch systems. Traditional kerosene (RP-1) or hypergolic propellant systems find an alternative in the combination of LOX and liquid methane (CH_4) as oxidizer and fuel, respectively. The LOX–CH_4 engine combination provides excellent performance alongside environmentally friendly characteristics and ISRU capabilities, which make it suitable for Earth-independent space missions [33]. Future space missions that require long-duration exploration of lunar and Martian terrain are increasingly using LOX–CH_4 engines because of their sustainable, high-performance propulsion [33]. Vacuum-specific impulse (*Isp*) for LOX–CH_4 engines falls between 360 and 380 seconds, thus placing them between LOX–hydrogen and LOX–kerosene engines [34]. LOX–CH_4 demonstrates superior propellant density compared to LOX–hydrogen, which leads to better volumetric performance while requiring simpler cryogenic storage solutions. The boiling point of CH_4 exceeds that of hydrogen by 91°C, which results in simpler insulation and tank requirements and superior long-term storage capabilities that are crucial for space and reusable launch vehicles. The combustion of LOX and CH_4 produces primarily water vapor and carbon dioxide, with no toxic or corrosive exhaust emissions. LOX–CH_4 engines provide a safer operational environment than hydrazine-based systems because they produce benign byproducts. CH_4 production from Martian CO_2 and water through the Sabatier reaction enables future on-site propellant synthesis for return missions and sustainable off-Earth operations [35]. The LOX–CH_4 system provides better engine reusability because it produces less coking and soot accumulation compared to RP-1 systems. Space companies SpaceX, Blue Origin, and ESA have funded LOX–CH_4 engine development to support crewed and robotic space missions with enhanced sustainability and reusability capabilities [74]. The strategic evolution of propulsion technology includes LOX–CH_4 engines, which provide favorable thermophysical properties with environmental responsibility and ISRU potential to support human and robotic exploration beyond Earth orbit into the future.

4.4 Hydrogen peroxide-based systems

The use of hydrogen peroxide (H_2O_2) or HTP as a monopropellant and oxidizer began in the 1930s when German rocket programs employed it during World War II. The early launch vehicles of Britain and the Soviet Union, along with experimental aircraft and torpedoes, utilized H_2O_2 as a propellant. Although its use declined with the rise of hydrazine and ammonium perchlorate systems, high-concentration H_2O_2 has regained attention as a "green propellant" due to its storability, catalytic decomposition into non-toxic byproducts (steam and oxygen), and potential for modern space propulsion. HTP-based propulsion systems represent one of the earliest and most versatile green propellant technologies, offering a viable alternative to toxic monopropellants such as hydrazine [36]. HTP serves as a monopropellant or as an oxidizer in bipropellant engines when combined with fuels like kerosene, alcohols, or hydrocarbons at high concentrations of 85%–98% by weight. Upon catalytic decomposition, HTP yields superheated steam and oxygen, which can generate thrust either directly or by combusting with a compatible fuel in a bipropellant configuration. In monopropellant applications, HTP is decomposed through contact with a catalyst bed, commonly silver or platinum, releasing gas at temperatures exceeding 600°C. Its moderate Isp in monopropellant form (150–180 seconds), together with its simplicity, rapid ignition, and low toxicity, makes it suitable for satellite attitude control, reaction control systems (RCS), and upper-stage operations [37]. In bipropellant systems, where HTP acts as an oxidizer, *Isp* can exceed 300 seconds depending on the fuel pairing, offering performance competitive with traditional chemical propulsion while significantly improving safety and environmental compliance [38]. HTP presents minimal risk to personnel and the environment because it is non-carcinogenic, non-persistent, and decomposes into benign byproducts, that is, water and oxygen. It also has relatively low vapor pressure and is easier to store and transport than many hypergolic propellants [39]. Concentrated HTP must be handled with care since it is chemically reactive and will decompose in the presence of contaminants, heat, or incompatible materials. The stability of storage and material compatibility remain essential factors in both design and operational procedures. Several space agencies and commercial entities have revisited HTP systems because of their green credentials and flight heritage [40]. Applications range from launch vehicle roll control systems to deep-space propulsion in monopropellant or bipropellant systems. **Figure 1a** presents a monopropellant system that decomposes a propellant through a catalyst bed to produce hot gases, generating thrust through a nozzle. In contrast, **Figure 1b** demonstrates a bipropellant system that uses separate fuel and oxidizer tanks to supply propellants into a combustion chamber for high-temperature gas generation and nozzle exhaust. The dual-use flexibility (monopropellant or bipropellant) and non-toxic profile of HTP, along with its simplicity, make HTP-based systems an attractive option for cost-effective and environmentally compliant propulsion architectures, particularly

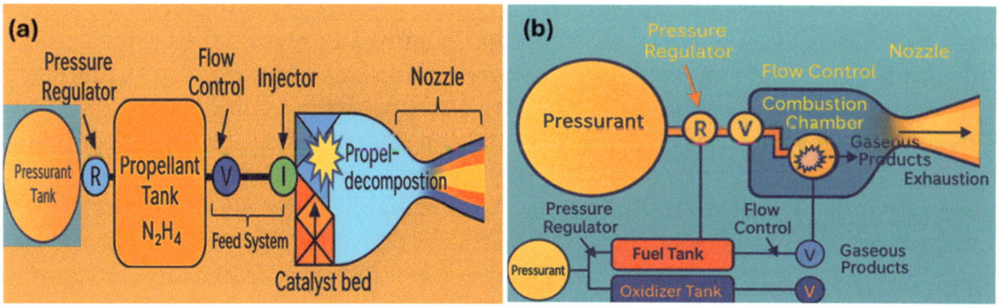

Figure 1.
Schematics of (a) monopropellant propulsion system and (b) bipropellant propulsion system; modified with permission from Alnaqbi et al. [20].

in small satellite platforms and responsive launch systems. In conclusion, HTP-based propulsion strikes a balance between heritage, environmental responsibility, and operational simplicity, positioning it as a practical green solution for both current and future aerospace applications.

5. Electric and hybrid propulsion as sustainable alternatives

Electric and hybrid propulsion systems serve as sustainable alternatives to traditional chemical propulsion because they provide better efficiency and reduced environmental impact, together with enhanced mission flexibility. Electric propulsion operates by using solar panels or onboard power system electrical energy to accelerate propellant at high velocities, which produces both higher specific impulse and better fuel efficiency compared to chemical engines [41]. The three main electric propulsion systems consist of ion thrusters, Hall-effect thrusters, and electrospray engines, which serve different mission requirements, including satellite station-keeping, orbital transfers, and deep-space exploration [42]. Hybrid propulsion systems unite chemical and electric propulsion elements to achieve optimal thrust performance together with extended operational duration. These technologies provide major operational and environmental benefits through their ability to reduce propellant weight and launch stresses and produce minimal toxic waste. Electric and hybrid propulsion systems will lead the development of next-generation spacecraft because space missions now focus on reusable systems, extended operational periods, and environmental sustainability.

5.1 Ion and hall-effect thrusters

The space industry depends on electric propulsion systems, including ion and Hall-effect thrusters, because these systems deliver high efficiency alongside extended operational lifespans for space missions involving orbit maintenance, deep-space exploration, and interplanetary transfers [43]. The principle behind these technologies involves electric or electromagnetic fields, which accelerate ionized propellant to achieve specific impulse levels beyond chemical propulsion systems, ranging from 1,500 to 3,000 seconds [44]. Ion thrusters use electrostatic acceleration to propel ions, usually xenon, through grids charged at high voltages. The spacecraft uses a neutralizer while accelerating ions to speeds greater than 30 km/s. This method delivers minimal thrust but maintains outstanding fuel efficiency, which suits space missions requiring prolonged, gradual acceleration, such as NASA's Vesta and Ceres exploration. **Figure 2a** presents the basic operational layout of the cathode in ion acceleration through structured grids in the gridded-ion system. The cathode maintains ion flow while establishing the base of this propulsion system, whereas the gridded-ion system, under high-power excitation, demonstrates its improved performance capabilities (see **Figure 2b**). It shows the role of energy input affecting thrust capability for advanced propulsion needs. Furthermore, the Hall thruster utilizes electric and magnetic fields to convert xenon propellant into ions, which then experience acceleration. The operation of Hall thrusters differs from ion thrusters since Hall thrusters do not need grids to function; instead, they use magnetic fields to confine electrons, which generates a Hall current that produces ionized propellant for electric field acceleration [44].

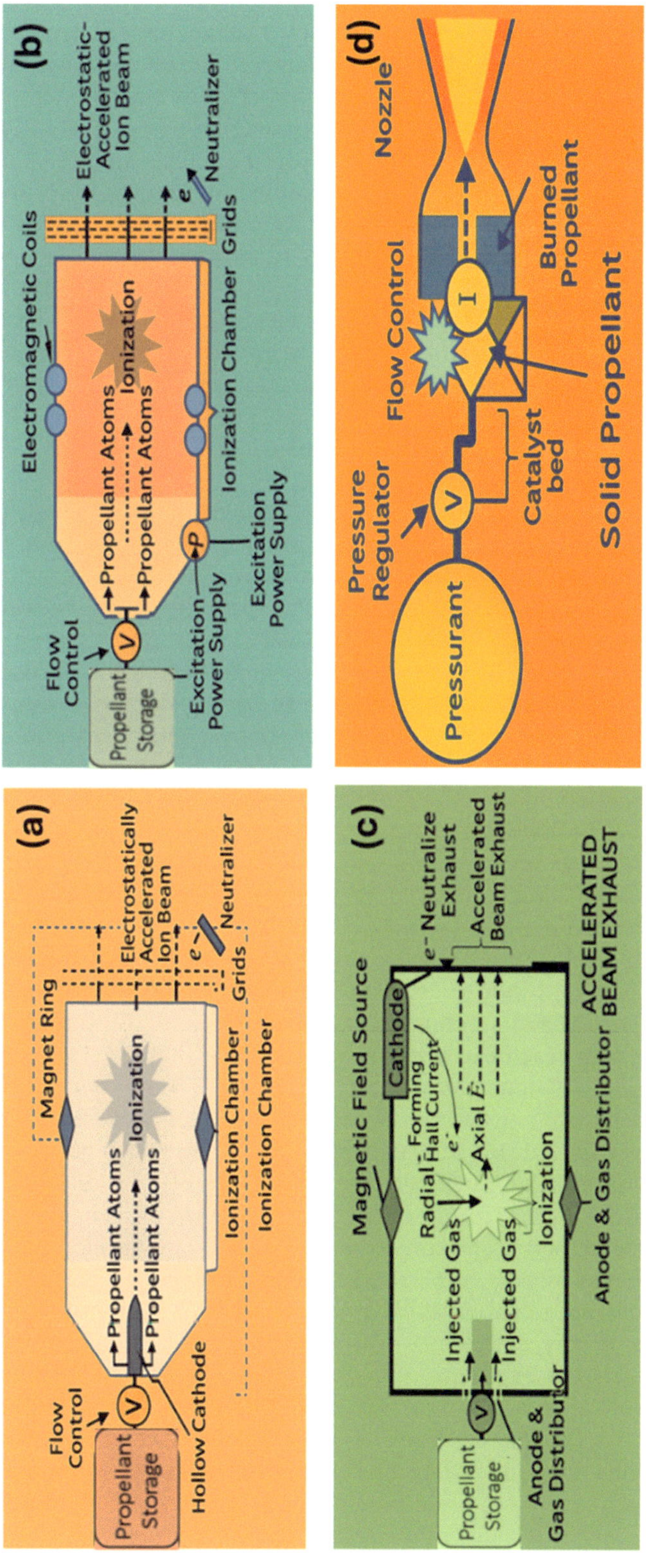

Figure 2.
(a and b) Gridded-ion system using cathode and high power excitation source, (c) schematic of hall propulsion, and (d) hybrid system configuration; modified with permission from Alnaqbi et al. [20].

Figure 2c schematically demonstrates the Hall propulsion system's ion acceleration through electric fields while using magnetic fields, achieving efficient continuous thrust operation through their basic system structure. The operation of Hall thrusters produces thrust densities that are higher than ion engines but have simpler mechanical designs and lower Isp performance, ranging from 1,500 to 2,000 seconds, which makes them suitable for commercial satellites and missions requiring moderate thrust levels. Both thruster types demonstrate significant propellant reduction capabilities compared to chemical systems, which leads to reduced launch mass together with lower emissions. The use of inert, non-toxic gas xenon enhances both environmental safety and compliance during storage and handling procedures. The rising need for sustainable space propulsion solutions drives the increasing adoption of ion and Hall-effect thrusters in both government and commercial space industries, supporting sustainable spacecraft design and mission planning.

5.2 Solar-electric and nuclear-electric options

Solar-electric propulsion (SEP) systems and nuclear-electric propulsion (NEP) systems operate as innovative space mission propulsion systems, providing sustained, efficient thrust for extended space missions through improved propellant efficiency and extended mission capabilities. The main difference between these technologies emerges from their method of power generation, as both use electrical thrusters such as ion and Hall-effect systems. SEP generates electrical power by using photovoltaic arrays to convert solar energy before driving electric thrusters [45]. SEP technology has established itself in satellite and interplanetary space missions operating within the inner solar system because of its ability to generate reliable power through solar irradiance. The specific impulse of SEP propulsion reaches 1,500–3,000 seconds, which leads to substantial propellant mass reductions, enabling improved payload delivery. BepiColombo, a joint ESA and Japanese Aerospace Exploration Agency (JAXA) mission to Mercury, and NASA's DART mission represent notable missions that use SEP technology [46]. SEP systems face two main limitations: their operation depends on sunlight, which stops functioning past the asteroid belt, and they require extensive and complex solar array deployment in space. The key difference between NEP and SEP is that NEP uses a miniaturized nuclear reactor to produce electricity, which provides power regardless of the Sun's distance (Mason). NEP systems offer major benefits to deep space and crewed missions that aim to reach outer planets because they operate without Sun-related limitations and sustain power for extensive periods. The combination of a robust high-power energy source with efficient electric propulsion technology within NEP systems allows for faster flight times and adaptable mission trajectories with decreased propellant weight (Mason). The development of NEP technology remains promising, yet scientists need to solve problems related to reactor miniaturization, radiation shielding requirements, safety protocols, and nuclear launch regulatory compliance. The implementation of SEP and NEP systems leads to sustainable space exploration through their capability to reduce propellant usage while decreasing launch weight and extending mission durations. Their scalable and modular architecture features make them ideal candidates for upcoming exploration missions, including Mars transport, asteroid rendezvous, and lunar logistics operations, because of their evolving propulsion requirements for efficiency, safety, and environmental responsibility.

5.3 Hybrid concepts and combined propulsion strategies

The development of spacecraft design focuses on hybrid propulsion concepts, which merge multiple propulsion technologies to fulfill various mission requirements. These systems use chemical propulsion together with electric and solar or nuclear energy sources to enhance performance during different mission phases, starting from launch and extending to orbital transfer and deep space maneuvering, with a focus on sustainability, operational flexibility, and efficiency. The hybrid system utilizes chemical propulsion with high thrust capability for initial orbit insertion and quick trajectory changes, followed by electric propulsion through ion or Hall-effect thrusters for precise orbit adjustments and deep space cruising [47]. Through staged propulsion methods, spacecraft gain the powerful capabilities of chemical systems together with electric propulsion efficiency, which decreases total propellant mass and makes missions more cost-effective. **Figure 2d** presents a hybrid system configuration, combining elements of gridded-ion and Hall propulsion technologies. The integration emphasizes a balance between efficiency and power density, showcasing the role of hybrid systems in addressing the limitations of standalone propulsion methods. The Psyche mission operated by NASA uses chemical propulsion for initial movement, yet relies on SEP to travel through interplanetary space. Advanced hybrid propulsion systems combine solar-thermal power with high-efficiency electric thrusters and nuclear-electric power generation to enable long-duration missions [48]. These systems function best for crewed exploration, cargo transport, and planetary surface access because they fulfill variable thrust requirements with reduced resupply needs. Government agencies, together with commercial providers, are actively developing satellite servicing capabilities using SEP systems with chemical kick stages and hybrid chemical-electric tugs. From a sustainability standpoint, hybrid systems offer significant advantages by reducing the dependence on high-toxicity propellants and minimizing the environmental footprint associated with launch mass and fuel consumption. These systems enable increased mission independence while providing reusability and adaptability features essential for developing modern space infrastructure networks. Hybrid propulsion methods aid sustainable space operations through practical solutions by merging various technologies to deliver high performance and environmental advantages.

5.4 Comparative analysis

Table 1 demonstrates a diverse range of chemical, electric, and hybrid advanced propulsion technologies currently used by developers and operators for their unique performance characteristics, operational advantages, and limitations. The less toxic green monopropellants HAN and ADN offer performance equivalent to hydrazine while providing 250–265-second specific impulse values and denser characteristics, leading to superior density impulse [49]. The storability and efficiency of AF-M315E and glycol blends as HAN-based formulations are good, but their thermal stability, ignition delays, and catalyst degradation pose operational challenges. The performance of ADN systems like LMP-103S and ADN–HAN mixtures is enhanced and less toxic, but their reduced storability and minimal flight history act as barriers to increased adoption [50]. LOX–CH_4 propulsion operates at high theoretical specific impulse values of 360–375 seconds because it burns fuel

Propellant	Propulsion system	Density (g/cm^3) @ 25°C	Critical pressure (MPa)	Critical temperature (K)	Theoretical *Isp* (seconds)	Advantages	Disadvantages	Reference
Hydroxylammonium nitrate (HAN)	Green monopropellant thruster	1.46	~7.5	~650	250–260	Low-toxicity alternative to hydrazine; storability; higher density impulse	Catalyst degradation; thermal stability concerns	[49]
HAN + Methanol/ Amine blend (AF-M315E)	Monopropellant	1.47	~8	~670	252–258	High-density impulse; lower toxicity than hydrazine	Higher combustion temperature, material stress	[57]
HAN + Glycol-based formulations	Monopropellant	1.45	~7	~640	~250	Safer handling, green certification	Limited heritage, ignition delay issues	[58]
Ammonium dinitramide (ADN) solution – LMP-103S	Monopropellant	1.24	~7	~720	250–260	Non-toxic; higher *Isp* than hydrazine	Lower storability; limited catalysts	[50]
ADN + Methanol/ Ethanol blends	Monopropellant	1.20–1.25	~7	~710	250–255	Good performance; less toxic	Limited flight heritage	[59]
ADN + HAN mixtures	Monopropellant	1.30	~7	~700	260–265	Enhanced performance	Thermal stability challenge	[60]
LOX–Methane	Cryogenic bipropellant	0.42 (LOX ~1.14, CH_4 ~0.42)	4.6 (CH_4), 5.2 (O_2)	190 (CH_4), 154 (O_2)	300–340	High *Isp*; ISRU potential on Mars; cleaner combustion versus RP-1	Cryogenic storage; boil-off issues	[51]
Hydrogen peroxide (90% H_2O_2)	Monopropellant/ bipropellant	1.40	~22	~740	160–180 (mono), 280–300 (bi with kerosene)	Storability: decomposes to O_2 + steam; relatively green	Handling hazards: catalyst coking	[61]
H_2O_2 + Kerosene	Bipropellant	1.05–1.10	~22	~740	300–325	Simple ignition; moderate performance	High decomposition sensitivity	[72]
Ion thruster (Xenon)	Electrostatic ion propulsion	Gas @1 atm ~0.0059	N/A	N/A	3,000–4,500	Extremely high *Isp*; efficient for deep space	Very low thrust; requires high power	[62]

DOI: http://dx.doi.org/10.5772/intechopen.1013180

Propellant	Propulsion system	Density (g/cm^3) @ 25°C	Critical pressure (MPa)	Critical temperature (K)	Theoretical *Isp* (seconds)	Advantages	Disadvantages	Reference
Hall-effect thruster (Xenon)	Plasma propulsion	Gas @1 atm ~0.0059	N/A	N/A	1,500–2000	Higher thrust density than ion thrusters; heritage on GEO sats	Erosion of channel walls; power-hungry	[52]
Krypton Hall-effect thruster	Plasma propulsion	Gas @1 atm ~0.0034	N/A	N/A	1,200–1800	Lower cost of propellant than xenon.	Lower performance than Xe	[53]
Solar-electric propulsion (SEP, Xenon)	Electric propulsion	~0.0059	N/A	N/A	2000–4,000	Scalable power: long-duration thrust	Limited by solar flux; mass of solar arrays	[63]
Nuclear-electric propulsion (NEP, Xenon)	Electric propulsion	~0.0059	N/A	N/A	3000–5,000	Constant power beyond solar distance; high endurance	Political/regulatory issues; reactor mass	[71]
Hybrid Propellant (H_2 O_2 + Rubber fuel grain)	Hybrid rocket	Solid + liquid mix (~0.9)	~22 (H_2 O_2)	~740	280–320	Simplicity: throttle/stop possible	Regression rate limits; efficiency issues	[54]
Hybrid (LOX + Hydrocarbon grain)	Hybrid rocket	0.9–1.0	LOX 5.2	LOX 154	~330	High performance potential	Combustion instability	[64]
Combined Chemical + Electric	Dual-mode spacecraft propulsion	Varies	Varies	Varies	~350 (chem) + 2000–4000 (electric)	Flexibility: high thrust for maneuvers + efficient cruise	Mass/complexity; integration challenges	[55]
Solar-Sail + Electric Hybrid	Hybrid propulsion	N/A	N/A	N/A	Electric: 2000–4000; Sail: infinite	Virtually unlimited Δv from sunlight	Very low thrust; requires a large sail	[56]

Table 1.
Characteristic features of various propellants and propulsion systems.

cleanly while allowing Mars ISRU capabilities [51]. The storage problems, along with boil-off losses and complex ground operations, affect these cryogenic propellants even when densified. Hydrogen peroxide systems act as both monopropellants and bipropellants while providing moderate performance (160–325 seconds) and storability advantages because of their density of 1.40 g/cm^3, but operational safety risks stem from decomposition sensitivity and catalyst degradation. Xenon [52] and krypton-based [53] ion thrusters and Hall-effect thrusters achieve extremely high *Isp* values between 1,200 and 4,000 seconds through their efficient operation. These systems excel for deep space missions and station-keeping duties, yet they face limitations from minimal thrust output, erosion problems, and high-power requirements, with krypton providing cost-effective operation at reduced performance levels. SEP extends efficiency for inner solar system missions, yet it faces restrictions from solar flux and array mass requirements, and NEP extends capability beyond solar distance with high endurance but encounters regulatory and mass-related challenges. The *Isp* range for hybrid propulsion systems that use liquid oxidizers with solid hydrocarbon-based fuels extends from 280 to 320 seconds [54]. The system design simplicity and throttle ability of these systems make them appealing, but regression rate constraints, efficiency problems, and combustion instabilities limit their effectiveness. The development of integrated propulsion concepts includes dual-mode chemical-electric systems, which use chemical propulsion for fast maneuvers and electric propulsion for cruising to achieve mission flexibility at the expense of additional weight and complexity [55]. The combination of solar sail technology with electric propulsion allows for unlimited △v capability from solar power but requires massive sail structures while providing precise trajectory control through electric thrust [56]. The table demonstrates that propulsion systems lack universal application because each system requires different efficiency levels, thrust capabilities, storage capabilities, operational risk tolerance, complexity levels, mission requirements, and exploration sustainability goals.

The assessment of spacecraft propulsion systems requires evaluation of specific impulse (*Isp*) against thrust-to-power ratio, as this relationship demonstrates propellant efficiency against thrust generation ability. The three fundamental propulsion technologies –namely, electric, kinetic, and chemical systems – exist as separate domains within the performance spectrum, as shown in **Figure 3a**. The *Isp*

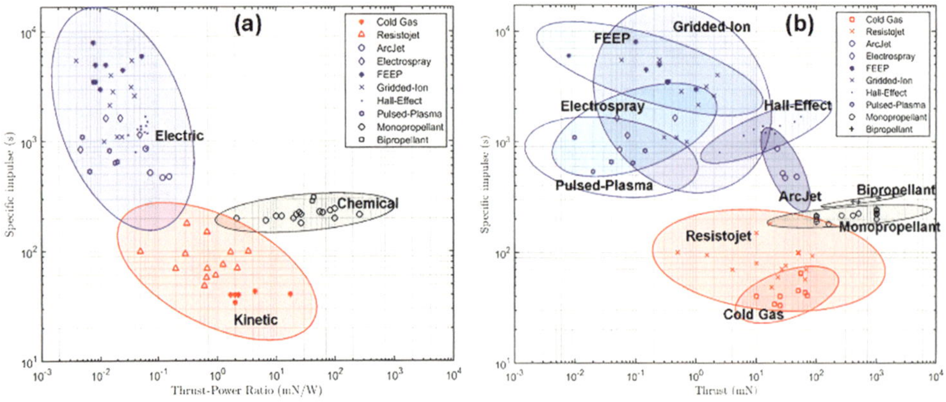

Figure 3.
(a) Thrust-to-power ratio and (b) Isp *performance of different propulsion technologies; with permission from Alnaqbi et al. [20].*

range of electric propulsion systems extends from about 1,000 seconds to almost 10,000 seconds across their technologies, including gridded-ion thrusters, Hall-effect thrusters, electrospray devices, field emission electric propulsion (FEEP), arc-jets, and pulsed plasma thrusters. These systems provide remarkable efficiency because they conserve propellant, which is essential for extended space missions, interplanetary transfers, and situations requiring minimal propellant mass. Electric propulsion systems achieve high efficiency but generate low thrust due to their thrust-to-power ratio (below 1 mN/W) and total thrust output, which reaches only 1 N [20]. Their inherent limitations prevent them from being used for fast orbital maneuvers, launch operations, or missions requiring quick payload repositioning. The specific impulse values of kinetic propulsion systems, which include cold-gas and resistojet thrusters, range between 50 and 200 seconds, while their thrust-to-power ratios exceed those of electric systems. These systems offer advantages due to their straightforward design, reliable operation, and fast response characteristics, making them ideal for attitude control, small satellite maneuvers, and missions requiring adaptability and robustness rather than fuel efficiency. Chemical propulsion systems, which include monopropellant and bipropellant thrusters, are positioned in the middle of the spectrum because they achieve 200–500 seconds of specific impulse and produce thrust-to-power ratios exceeding 1 mN/W [20]. These systems generate thrust levels ranging from tens to thousands of millinewtons, making them vital for orbital insertion, rapid trajectory corrections, and launch vehicle operations that require both strong thrust and acceptable efficiency. **Figure 3b** illustrates the spacecraft propulsion system performance characteristics, linking *Isp* measurements of propellant efficiency with thrust force generation. The three propulsion domains create distinct areas in the thrust-to-*Isp* map, which shows electric systems in the high-Isp but low-thrust sector, chemical systems between moderate efficiency and substantial thrust, and kinetic systems in the low-to-moderate *Isp* segment with better thrust responsiveness than electric thrusters. The design trade-offs of propulsion systems are evident as electric systems optimize efficiency but reduce thrust output; chemical systems strike a balance between efficiency and thrust performance for mission versatility; and kinetic systems prioritize reliability and responsiveness over performance efficiency. The selection of propulsion systems requires matching mission needs with payload restrictions and operational requirements. Current developments focus on creating hybrid propulsion systems that combine beneficial aspects from different domains, utilizing electric propulsion for efficient cruise phases and chemical propulsion for high-thrust maneuvers. The integration of different propulsion systems enables enhanced mission adaptability and offers better prospects for sustainable space exploration through improved resource management and optimized system operation.

6. ISRU and green fuel production

The combination of ISRU with green fuel production enables sustainable space exploration through the direct extraction of propellants and life-support resources from extraterrestrial environments. The main objective of ISRU technologies involves extracting water ice, carbon dioxide, and regolith from lunar, Martian, or asteroidal environments to create usable propellants, including LOX, CH_4, and hydrogen [65]. The technology decreases mission expenses and launch weight while enabling extended space missions with crew members through improved self-sufficiency and operational stability. The implementation of ISRU for green fuel production supports environmental sustainability and creates closed-loop systems, making planetary colonization more feasible. Space agencies, together with commercial ventures, are making sustainability and extended Earth presence a priority through ISRU, which enables propulsion strategies when combined with green engines that use non-toxic, regenerative propellants from local resources.

6.1 Propellant production on Mars and the Moon

The production of propellants on Mars and the Moon serves as a fundamental part of ISRU because it enables both sustainable and cost-effective space exploration. The two celestial bodies provide distinct opportunities to produce essential propellants, including oxygen, hydrogen, and CH_4 through local material processing, which reduces Earth-based resupply needs and decreases mission expenses [24]. The Sabatier reaction enables the conversion of carbon dioxide (CO_2) in the Mars atmosphere into CH_4 and oxygen (O_2) when hydrogen is present [66]. The extraction of Martian water ice through heating or chemical processing of regolith provides the necessary hydrogen for this process. NASA's MOXIE (Mars Oxygen ISRU Experiment) validated the CH_4–oxygen fuel combination, which works perfectly for liquid bipropellant engines, by successfully producing oxygen from CO_2 on Mars [66]. The Moon lacks an atmosphere, which creates different opportunities for ISRU. The oxygen content in lunar regolith reaches 45% by weight in oxide form. Modern technologies include ilmenite reduction through $FeTiO_3$ processing as well as molten electrolysis to extract oxygen [67]. The Moon lacks native CH_4 deposits, but local oxygen production could be used to combine imported or synthesized fuels like hydrogen or CH_4 for hybrid chemical propulsion systems. The production of propellants on-site provides enhanced mission flexibility and enables both return missions and extended surface operations. These advancements, together with green propulsion systems, create the foundation for a sustainable interplanetary infrastructure.

6.2 Integration with mission architecture

The implementation of green propellant technologies, together with ISRU systems, enables sustainable mission architectures that support extended space exploration. Modern mission planning has moved beyond independent propulsion system design to create integrated architectures that unite payload restrictions with launch strategies, environmental standards, risk reduction methods, and extended operational goals [68]. **Figure 4** illustrates the role of ISRU in enabling sustainable propellant production for lunar and Martian missions. **Figure 4(a)** presents the ISRU propellant production flow, beginning with local feedstocks such as water ice, carbon dioxide, and regolith. Through extraction, electrolysis, and Sabatier reactions, these raw materials are converted into usable propellants, including LOX, CH_4, and hydrogen (H_2). Green propulsion systems simplify satellite servicing operations and repositioning activities, as well as deorbiting procedures, while decreasing the logistical challenges that come with dangerous propellants in both LEO and GEO orbit regions [69]. **Figure 4(b)** demonstrates the embedment of ISRU-derived propellants within a broader mission architecture. Green propellants,

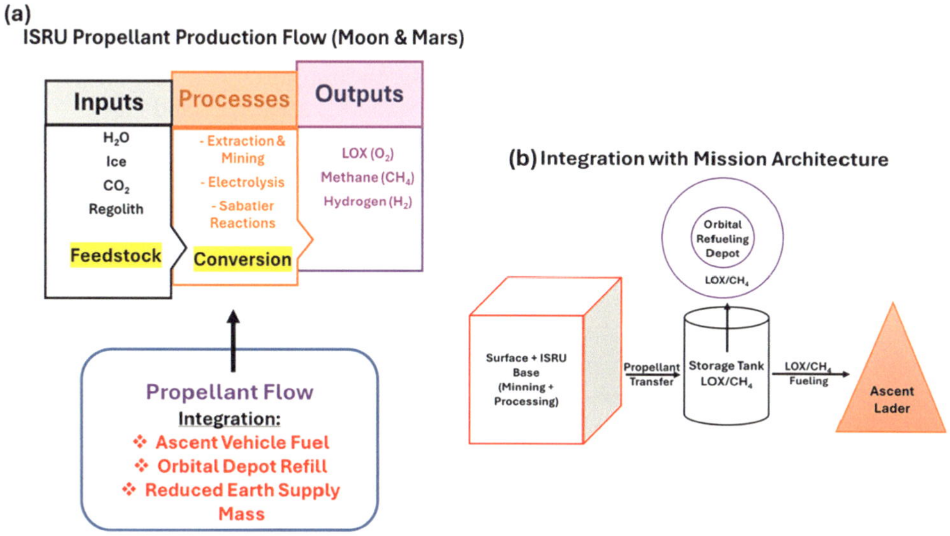

Figure 4.
In situ *resource utilization (ISRU) for the Moon and Mars enables conversion of local feedstocks (H_2O ice, CO_2, regolith) into propellants such as LOX, CH_4, and H_2. (a) Propellant production flow showing feedstock, conversion processes, and outputs; and (b) integration with mission architecture, where ISRU-derived propellants support ascent vehicles, orbital depots, and reduce Earth resupply needs.*

including LOX and CH_4, serve as standard propulsion elements for deep-space missions aiming to reach the Moon and Mars because they work well with ISRU capabilities. Propellant generated at the surface of the ISRU base is transferred to storage tanks containing LOX/CH_4 and can then be supplied to orbital depots or ascent vehicles. This architecture illustrates the operational flexibility of ISRU, ensuring that local production surface missions are less dependent on Earth-based resupply while simultaneously enabling orbital logistics, vehicle refueling, and long-duration exploration campaigns. The integration of electric propulsion systems, using solar or nuclear power, becomes natural for mission architectures focusing on high-efficiency transfer orbits, station-keeping operations, and multi-target exploration. The combination of chemical and electric systems through hybrid approaches powers space tugs, orbital transfer vehicles, and landers to meet their high-thrust needs while maintaining long-duration efficiency [70]. Together, the two panels highlight the linkages of ISRU resource extraction, chemical conversion, and mission integration, positioning it as a cornerstone of sustainable lunar and Martian exploration strategies. The integration of this conversion process directly supports multiple mission requirements, such as ascent vehicle fueling, orbital depot refilling, and reduction of Earth-supplied propellant mass, thereby decreasing launch costs and logistics complexity. The current mission design approach incorporates sustainability metrics that evaluate propellant life-cycle impact along with launch mass optimization and reusability features. The environmental advantages of green propellants enable faster launch facility operations and expanded mission authorizations through environmental regulations. Green propulsion, together with ISRU-based technology, has evolved from innovation to standard practice for designing satellite constellations and crewed interplanetary missions, resulting in safer operations with cleaner methods and autonomous control of the developing space economy.

6.3 Technical and economic feasibility

The implementation of green propulsion systems and ISRU technologies depends on their technical maturity, combined with operational reliability and cost-effectiveness relative to conventional propellants and infrastructure. The technical maturity of green propellants AF-M315E, LMP-103S, and HTP is high because multiple successful in-orbit tests have proven their operational safety. These alternative propulsion systems match or surpass the performance of hydrazine-based systems while providing reduced handling risks and toxicities. The reduced need for expensive safety infrastructure and specialized personnel makes launch and ground-handling facilities more operationally efficient. Commercial and scientific missions successfully utilize electric propulsion systems with ion and Hall-effect thrusters. Solar or nuclear power sources enable these propulsion systems to achieve exceptional fuel efficiency but provide limited thrust capabilities for orbital station-keeping and interplanetary cruise applications. The economic feasibility of integrating these systems has increased because launch costs decrease when combined with reduced launch mass. The economic potential of ISRU technology stands strong, even though it requires more complex technical development. The production of propellant from lunar or Martian resources eliminates the need to transport substantial Earth-based fuel masses, thus lowering both mission weight and expenses. The current ISRU technologies exist at lower TRLs, yet advancements in autonomous robotics, combined with thermal processing and chemical conversion technologies, are speeding up their practical implementation. NASA's MOXIE experiment has verified the ability to produce oxygen by processing Martian CO_2. The implementation of green propulsion technology reduces mission expenses through simpler logistics, fewer regulatory requirements, and shorter turnaround periods. The combination of ISRU with green propulsion technology enhances long-term cost efficiency, primarily for missions that involve multi-platform designs, reusable spacecraft, and permanent planetary bases. The combined technical reliability of green and ISRU-based propulsion systems, together with strategic cost advantages, makes them increasingly viable options for sustainable space exploration despite initial research and development (R&D) expenses for infrastructure development.

7. Challenges, trade-offs, and future research directions

7.1 Material compatibility and storage issues

The adoption of green propellants faces essential hurdles because material compatibility and secure long-term storage remain vital for both propulsion system reliability and mission safety. The chemical characteristics of green propellants produce unpredictable material interactions with conventional storage and system components because these propellants lack the toxicity of legacy chemicals like hydrazine. The HAN- and ADN-based propellant mixtures show high reactivity because they exist as water-based solutions that react unfavorably when exposed to contaminants or incompatible metals, including aluminum and copper. System integrity and mission performance become endangered, as these interactions result in corrosion damage to components and degraded sealing materials, as well as catalytic decomposition processes. Long-term validation of advanced materials, including passivated stainless steel, titanium alloys, and specially coated composites, remains necessary to establish confidence throughout mission lifecycles since these materials are being studied to reduce associated risks. Storage challenges are also non-trivial. Most green propellants show hygroscopic behavior while also being thermally sensitive, thus requiring stable storage environments for performance maintenance and chemical stability. The storage of HAN and LMP-103S requires precise temperature control to prevent phase separation and decomposition over time. HTP exists as an environmentally friendly fuel that decomposes explosively when contaminated or exposed to heat and light, so storage requires clean, temperature-controlled facilities with specific handling procedures. System design requires trade-offs between performance capabilities and both system complexity and cost levels. Although green propellants decrease environmental risks and regulatory compliance burdens, they might need stronger materials and customized insulation and thermal control systems, which reduce some logistical benefits. The development of corrosion-resistant materials, advanced coatings, and additive manufacturing methods represents a future research priority for green propellant compatibility. Long-duration compatibility and aging tests under simulated space conditions should be standardized to validate storage solutions and materials for extended missions, particularly for planetary exploration and reusable spacecraft operations.

7.2 Scalability and system readiness levels

The adoption of green propulsion technologies faces two major barriers due to their limited scalability and system readiness levels (SRLs) across different mission types and scales. The green propellants AF-M315E and LMP-103S have achieved TRLs between 6 and 7 because they have been successfully tested in operational and relevant environments. However, the implementation of these propellants remains restricted for large-scale and high-thrust applications, especially in heavy launch vehicles, human-rated systems, and interplanetary platforms. Scalability problems arise during both the manufacturing operations and the integration of propulsion systems. The commercial production of green propellants needs to maintain consistent purity and stability throughout large batches via certified supply chains and strict quality control measures, which the industry still needs to develop. The thermal and flow properties, along with the ignition characteristics of green propellants, require multiple modifications to existing propulsion systems designed for traditional fuels. The deployment schedule becomes longer, and initial development expenses increase when using green propellants for missions that require high thrust levels or extended operational periods. Electric propulsion systems operate at scale for low-thrust applications yet encounter similar barriers when attempting high-power implementations. Hall-effect and ion thrusters demonstrate excellent performance in satellite and robotic exploration, but they cannot achieve rapid orbit transfers or meet crewed mission requirements for higher thrust levels. Future research needs to focus on developing power scalability, propulsion modularity, and component standardization to overcome current limitations. The deployment of green propulsion across diverse mission classes requires ongoing flight demonstrations, along with mission-level validation and production infrastructure investments, to advance SRLs.

7.3 Emerging research areas and innovation opportunities

The shift toward sustainable propulsion in space exploration drives innovation across different fields, creating new research opportunities and technological development possibilities. The maturation of green propellants has led researchers to focus on optimizing entire systems through propellant synthesis, advanced materials development, engine design, and autonomous handling. The focus of researchers today is to create innovative green propellant chemistries that surpass HAN and ADN. Scientists have investigated three promising propellant alternatives, which include ionic liquids, gelled propellants, and bio-derived fuels, due to their enhanced stability properties, non-volatile characteristics, and ISRU compatibility. These compounds show potential for customization to match the requirements of specific space missions, such as storage for deep-space probes or ignition at lunar low temperatures. The field of additive manufacturing for propulsion components remains an active area of research. The three-dimensional (3D) printing of thruster parts using corrosion-resistant alloys or ceramic composites enables engineers to create optimized geometries while reducing mass and speeding up prototyping, which becomes essential for engine adaptations to green fuel flow and combustion properties. The development of electric propulsion depends on improved power electronics efficiency, plasma diagnostics, and magnetically shielded thruster designs, which enhance thrust-to-power ratios and operational longevity. These technological advancements are particularly important for space missions that use solar-electric or nuclear-electric power systems. The development of automated ground support systems with intelligent features represents a substantial innovation opportunity. The safer handling capabilities of green propellants enable the implementation of automated fueling systems with leak detection and health monitoring capabilities, streamlining operations and supporting fast launch cycles. ISRU-linked research now focuses on developing large-scale reactors to produce oxygen, CH_4, and hydrogen on the Moon and Mars. The establishment of a closed-loop infrastructure for long-duration missions depends on integrating energy storage with cryogenic liquefaction and on-site propulsion system refueling. Furthermore, the development of novel catalytic materials may also enhance the performance of green propellants, specifically HTP, and must be extensively explored. The combination of these innovative areas transforms propulsion systems into safer, cleaner, and more versatile systems, supporting long-term sustainability goals in space exploration.

7.4 Implications for sustainable space exploration

The combination of green propulsion technologies with ISRU strategies creates substantial effects for sustainable space exploration in the future. The expanding scale and complexity of space missions, along with their increasing duration, make conventional propellant systems more difficult to sustain. Green propulsion systems tackle these problems head-on through their ability to decrease toxic substances while reducing operational risks and environmental effects both on Earth and in space. These technologies allow for safer ground operations and faster launch preparations, and they meet the requirements of tightening international environmental regulations, which are essential for frequent and long-term missions. The implementation of ISRU technology enables propellant production on the Moon,

Mars, and other celestial bodies, which reduces the need for expensive Earth-based resupply operations. This capability enables autonomous mission operations, reusable systems, and permanent human exploration infrastructure development. The combination of green propellants with ISRU-derived resources (e.g., LOX–CH_4) enables the development of a regenerative propulsion ecosystem that establishes the foundation for interplanetary logistics networks, surface base operations, and return missions. The transition to sustainable propulsion systems helps the space sector meet global climate and ESG goals while promoting responsible space utilization as a shared domain. The implementation of sustainable propulsion systems enables the creation of cleaner launch systems, environmentally neutral orbital operations, and planetary protection principles by minimizing contamination risks. The adoption of green technologies by commercial entities leads to better market competitiveness, reduced insurance expenses, regulatory costs, and access to government-backed sustainability programs. Green propulsion systems with ISRU capabilities serve as essential foundational elements for establishing sustainable exploration that can scale and maintain ethical standards throughout extended missions. These systems will become vital for human infrastructure expansion in space because they will ensure exploration remains ambitious while being sustainable, safe, and economically viable.

7.5 Policy and industry recommendations

The adoption of green and sustainable propulsion technologies requires coordinated efforts between policymakers, regulators, and industry stakeholders to update safety and certification frameworks for low-toxicity propellants and standardize international operations. Public funding for applied R&D expansion, combined with public–private partnerships and ISRU-related innovation support, will connect laboratory research to commercial deployment. Flight demonstration programs, together with robust qualification pipelines, will validate performance in operational environments, build market confidence, and foster trust. The development of certified, reliable green propellant supply chains, along with compatible infrastructure investments, will reduce integration expenses, shorten delays, and enhance worldwide accessibility. Mission planning guidelines that include sustainable propulsion considerations will drive early adoption, while ESG framework alignment enables operators to demonstrate environmental responsibility, achieve better regulatory compliance, and access green financing. The combination of these measures will establish sustainable propulsion as a fundamental element of future space missions, ensuring both scalability and long-term viability.

8. Conclusion

The chapter demonstrates the green and alternative propellants enabling sustainability in space exploration. Traditional chemical propulsion provides reliable performance, yet it produces major environmental problems, as well as safety and regulatory issues, because it requires toxic substances and dangerous operations. The propellant alternatives HAN, ADN, HTP, and LOX–CH_4 achieve technical readiness through flight testing while providing improved performance through cleaner operations and higher efficiency. Electric and hybrid propulsion systems improve interplanetary and long-duration mission capabilities through better fuel efficiency and flexibility, while ISRU on the Moon or Mars reduces the requirement for Earth-based propellant supply chains. The development of propellant chemistry, together with additive manufacturing and automated handling systems, advances these technologies, but material compatibility issues, scalability problems, and system readiness challenges continue to exist. These technologies need policy reforms, together with regulatory alignment and industry collaboration, to achieve widespread adoption. The innovations establish a foundation for upcoming space infrastructure, which will emphasize safety, autonomy, and environmental sustainability through green propulsion and ISRU systems for responsible exploration during the next decades.

Acknowledgement

The authors would like to thank the United Arab Emirates (UAE) Space Agency and Khalifa University Space Innovation and Technology Laboratory (KUSTIL) for funding the project and providing the resources.

Author details

Abdul Hai[1,2]*, Djamal Darfilal[1]*, Fatima AlHammadi[1], Salma S. Syed[3], Hamed Alhashmi[4], Abderrahim Nabi[1] and Sean Shan Min Swei[1]*

1 Khalifa University Space Technology & Innovation Laboratory (KUSTIL), Department of Aerospace Engineering, Khalifa University of Science and Technology, Abu Dhabi, United Arab Emirates

2 Department of Chemical & Petroleum Engineering, Khalifa University of Science & Technology, Abu Dhabi, United Arab Emirates

3 Department of Chemistry, Khalifa University of Science and Technology, Abu Dhabi, United Arab Emirates

4 Space Projects Management Section, United Arab Emirates Space Agency, Abu Dhabi, United Arab Emirates

*Address all correspondence to: abdul.hai@ku.ac.ae; djamal.darfilal@ku.ac.ae; sean.swei@ku.ac.ae

References

[1] Blondel-Canepari L, Sarritzu A, Pasini A. A holistic approach for efficient greener in-space propulsion. Acta Astronautica. 2024;**223**:435–447. DOI: 10.1016/j.actaastro.2024.07.023

[2] Mendoza PCA, Gonçalves RFB, Gouvêa LH, Pereira LGF. Composition optimization of a hypergolic green propellant based on monoethanolamine, n-butanol and 90% hydrogen peroxide. Acta Astronautica. 2025;**229**:140–148. DOI: 10.1016/j.actaastro.2024.12.059

[3] Dallas JA, Raval S, Alvarez Gaitan JP, et al. The environmental impact of emissions from space launches: A comprehensive review. Journal of Cleaner Production. 2020;**255**:120209. DOI: 10.1016/j.jclepro.2020.120209

[4] Olawade DB, Ijiwade JO, Wada OZ. Toward net-zero in space exploration: A review of technological and policy pathways for sustainable space activities. Science of The Total Environment. 2025;**972**:179145. DOI: 10.1016/j.scitotenv.2025.179145

[5] ESA. ESA Green Agenda: Strategy and actions for overcoming climate and sustainability challenges. Paris, France: European Space Agency. 2023

[6] Groen FJ, Nguyen JT, Benardini JN. Responsible exploration - preserving the cosmos for tomorrow. Type of publication: NASA Technical Memorandum, Document ID: 20240007653. Publisher: National Aeronautics and Space Administration; 2024

[7] Maggi F, Carlotti S, Kappenstein C. Environmental impact of propulsion systems and green alternatives. In: Sgobba T, Musgrave GE, Johnson G, Kezirian MT. Safety Design for Space Systems. London, UK: Elsevier. 2023. p. 513–551

[8] Kasbi Y, Remissa I, Toshtay K, Mabrouk A, Bachar A, Azat S, Nosseir AES, Tiwari A, Sabbar EM, Amrousse R. H O and HAN green monopropellants—A state-of-the-art review on their recent development, corresponding synthesized catalysts, and their possible use as thrusters. Catalysts. 2025;**15**:183

[9] Remissa I, Jabri H, Hairch Y, et al. Propulsion systems, propellants, green propulsion subsystems and their applications: A Review. Eurasian Chemico-Technological Journal. 2023;**25**:3–19

[10] Klem MD. LOX/Methane In-Space Propulsion Systems Technology Status and Gaps. Texas: El Paso; 2017

[11] Okninski A. Solid rocket propulsion technology for de-orbiting spacecraft. Chinese Journal of Aeronautics. 2022;**35**:128–154. DOI: 10.1016/j.cja.2021.07.038

[12] Kakami A, Kuranaga A, Yano Y. Premixing-type liquefied gas bipropellant thruster using nitrous oxide/dimethyl ether. Aerospace Science and Technology. 2019;**94**:105351. DOI: 10.1016/j.ast.2019.105351

[13] Davis SM, Yilmaz N. Advances in hypergolic propellants: Ignition, hydrazine, and hydrogen peroxide research. Advances in Aerospace Engineering. 2014;**2014**:1–9. DOI: 10.1155/2014/729313

[14] Hasan D, Grinstein D, Kuznetsov A, et al. Green Comparable Alternatives of Hydrazines-Based Monopropellant and Bipropellant Rocket Systems. London, UK: IntechOpen; 2019

[15] Carlotti S, Maggi F. Evaluating new liquid storable bipropellants: Safety and performance assessments. Aerospace. 2022;**9**:561. DOI: 10.3390/aerospace9100561

[16] Kipanga KB, Hassan M. Space activities for sustainability and the role of the national space agencies. Beijing Law Review. 2025;**16**:1227–1259. DOI: 10.4236/blr.2025.162063

[17] Ordonez Valles L, Blondel CL, Apel U, et al. Challenges and opportunities of green propellants and electric pump feeding for future european kick stages. Aerotecnica Missili & Spazio. 2022;**101**:227–241

[18] Aggarwal R, Patel IK, Sharma PB. Green propellant : A study. International Journal of Latest Trends in Engineering and Technology. 2015;**6**:83–87

[19] Nosseir AES, Cervone A, Pasini A. Review of state-of-the-art green monopropellants: For propulsion systems analysts and designers. Aerospace. 2021;**8**:20. DOI: 10.3390/aerospace8010020

[20] Alnaqbi S, Darfilal D, Swei SSM. Propulsion technologies for CubeSats: Review. Aerospace. 2024;**11**:502. DOI: 10.3390/aerospace11070502

[21] Kavipriya JV, Chavan B, Manmeetkaur AB. Comparative study on green monopropellants for rocket engines. EDUZONE International Peer Reviewed Multidisciplinary Journal. 2022;**11**(1):17

[22] Florczuk W, Rarata GP. (2017). Performance evaluation of the hypergolic green propellants based on the HTP for a future next generation spacecrafts. 53rd AIAA/SAE/ASEE Joint Propulsion Conference. American Institute of Aeronautics and Astronautics, Reston, Virginia

[23] Wu W, Shen J, Kong H, et al. Energy system and resource utilization in space: A state-of-the-art review. The Innovation Energy. 2024;**1**:100029. DOI: 10.59717/j.xinn-energy.2024.100029

[24] Amaya B, Ampani P, Arevalo E, et al. SHELL System for Heavy-Lift End-To-End Lunar Logistics. College Park, MD, USA: University of Maryland. 2023

[25] EDA "European Defence Agency". (2016). The Impact of Reach and Clp European Chemical Regulations on the Defence Sector. 1–311

[26] OSHA. Hazard Communication. Hazard Classification Guidance for Manufacturers, Importers, and Employers. Occupational Safety and Health Administration, U.S. Department of Labor 2016;**3844**:1–432

[27] Spores RA, Masse R, Kimbrel S, Mclean C. GPIM AF-M315E Propulsion System. In: AIAA/ASME/SAE/ASEE Joint Propulsion Conference; 28–30 Cleveland, Cleveland, OH, USA 2014; p. 1–12 [Accessed: 2020-December-16]

[28] Chen F, Xuan C, Lu Q, et al. A review on the high energy oxidizer ammonium dinitramide: Its synthesis, thermal decomposition, hygroscopicity, and application in energetic materials. Defence Technology. 2023;**19**: 163–195

[29] Anflo K, Möllerberg R. Flight demonstration of new thruster and green propellant technology on the PRISMA satellite. Acta Astronautica. 2009;**65**:1238–1249. DOI: 10.1016/j.actaastro.2009.03.056

[30] Kore R, Vashishtha A. Combustion Behaviour of ADN-Based Green Solid Propellant with Metal Additives: A Comprehensive Review and Discussion. Aerospace. 2025;**12**:46. DOI: 10.3390/aerospace12010046

[31] Anflo K, Crowe B. (2011). In-Space Demonstration of an ADN-based Propulsion System. 47th AIAA/ASME/SAE/ASEE Joint Propulsion Conference & Exhibit. American Institute of Aeronautics and Astronautics; Reston, Virigina

[32] Pozzoni L, Paravan C, Galfetti L (2022). ADN-based Formulations Analysis for Green Solid Propellants. Proceedings of the 9th European Conference for Aerospace Sciences. DOI: 10.13009/EUCASS2022-4909

[33] Boué Y, Vinet P, Magniant S, et al LOX/methane reusable rocket propulsion at reach with large scale demonstrators tested. Acta Astronautica. 2018;**152**:542–556

[34] Tiago J, Pinto R. (2019). A Survey of Modern and Future Space Propulsion Methods

[35] Kim YC. Exploring water production dynamics in a Sabatier reactor: A comprehensive experimental investigation. Fuel. 2025;**396**:135298. DOI: 10.1016/j.fuel.2025.135298

[36] Kopacz W, Okninski A, Kasztankiewicz A, et al. Hydrogen peroxide – A promising oxidizer for rocket propulsion and its application in solid rocket propellants. FirePhysChem. 2022;**2**:56–66

[37] Kokal U, Baysal M, Emerce NB, et al. Evaluation and Performance Optimization of a Hydrogen Peroxide-Based Green Monopropellant Thruster for Steady-State Operations. Aerospace. 2025;**12**:136

[38] Okninski A, Kindracki J, Wolanski P. Multidisciplinary optimisation of bipropellant rocket engines using H2O2 as oxidiser. Aerospace Science and Technology. 2018;**82–83**:284–293. DOI: 10.1016/j.ast.2018.08.036

[39] Moser D (2001) High-performance, non-toxic spacecraft propulsion system development. 37th Joint Propulsion Conference and Exhibit. American Institute of Aeronautics and Astronautics: Reston, Virigina

[40] Łukasiewicz – Institute of Aviation. World's first space platforms utilising H2O2 in concentration above 98% – next level of green space propulsion. SpaceNews; 2023. Available from: https://spacenews.com/worlds-first-space-platforms-utilising-h2o2-in-concentration-above-98-next-level-of-green-space-propulsion/

[41] Boretti A. A narrative review of solar electric propulsion for space missions: Technological progress, market opportunities, geopolitical considerations, and safety challenges. Journal of Space Safety Engineering. 2025. DOI: 10.1016/j.jsse.2025.07.004

[42] Hesla E. Electric propulsion. IEEE Industry Applications Magazine. 2009;**15**:10–13. DOI: 10.1109/MIAS.2009.932594

[43] Wagh SA, Pawar P. Challenges and innovations in electric propulsion systems for space transportation: A comprehensive review. Acceleron Aerospace Journal. 2025;**4**:929–944. DOI: 10.61359/11.2106-2511

[44] Levchenko I, Xu S, Teel G, et al. Recent progress and perspectives of space electric propulsion systems based on smart nanomaterials. Nature Communications. 2018;**9**:879

[45] Casanova-Álvarez M, Navarro-Medina F, Tommasini D. Feasibility study of a Solar Electric Propulsion mission to Mars. Acta Astronautica. 2024;**216**:129–142. DOI: 10.1016/j.actaastro.2024.01.001

[46] Benkhoff J, Murakami G, Baumjohann W, et al. BepiColombo-Mission overview and science goals. Space Science Reviews. 2021;**217**:90

[47] Mani KV. Combined chemical–electric propulsion design and hybrid trajectories for stand-alone deep-space CubeSats. 2020;**56**(6):1816–1830

[48] Genovese A, Maraqten N. Advanced electric propulsion concepts for fast missions to the outer solar system and beyond. Journal of the British Interplanetary Society. 2023;**76**:114–121. DOI: 10.59332/jbis-076-04-0114

[49] NASA Glenn Research Center. Green Propellant Infusion Mission (GPIM) Overview. Washington, DC, USA: National Aeronautics and Space Administration. 2019

[50] Thomas JC, et al. Combustion characterization of LMP-103S in strand burner experiments. In: 9th European Conference for Aeronautics and Space Sciences. EUCASS; 2022. DOI: 10.13009/EUCASS2022-4932

[51] Hernandez I, Orozco S, Hossain MA, Choudhuri A. Thermal performance analysis of LOX/LCH4 engine feed systems using CFD modeling. Fluids. 2025;**10**:62. DOI: 10.3390/fluids10030062

[52] Goebel DM, Katz I. Fundamentals of Electric Propulsion. Ion and Hall Thrusters. Hoboken, NJ, USA: John Wiley & Sons; 2008

[53] Nakles MR, Hargus WA, Delgado JJ, Corey RL. (2011). A performance comparison of xenon and krypton propellant on an SPT-100 hall thruster. The 32nd International Electric Propulsion Conference. Iepc-2011 003

[54] Wei -S-S, Li M-C, Lai A, et al. A review of recent developments in hybrid rocket propulsion and its applications. Aerospace. 2024;**11**:739

[55] Mani KV. Combined chemical-electric propulsion design and hybrid trajectories for stand-alone deep-space CubeSats. Journal of Spacecraft and Rockets. 2020b;**56**:1816–1830

[56] Ceriotti M, McInnes CR. Hybrid solar sail and solar electric propulsion for novel Earth observation missions. Acta Astronautica. 2011;**69**:809–821. DOI: 10.1016/j.actaastro.2011.06.007

[57] Masse RK, Allen M, Driscoll E, Spores RA. AF-M315E propulsion system advances and improvements. National Aeronautics and Space Administration; 2016. Available from: https://ntrs.nasa.gov/api/citations/20170001286/downloads/20170001286.pdf

[58] Kasbi Y, Remissa I, Toshtay K, et al. H2O2 and HAN Green Monopropellants —A State-of-the-Art Review on Their Recent Development, Corresponding Synthesized Catalysts, and Their Possible Use as Thrusters. Catalysts. 2025;**15**:183

[59] Anflo K, Persson S, Thormahlen P, Bergman G, Hasanof T. Flight Demonstration of an ADN-Based Propulsion System on the PRISMA Satellite. American Institute of Aeronautics and Astronautics; 2012. DOI: 10.2514/6.2006-5212

[60] Wingborg N, et al. (2006) Formulation and Characterization of ADN-Based Liquid Monopropellants. JANNAF Meeting Proceedings

[61] Cervone A, Torre L, D'Agostino L, et al. Development of hydrogen peroxide monopropellant rockets. Collected Technical Papers - AIAA/ ASME/SAE/ASEE 42nd Joint Propulsion Conference. 2006;**11**:8786–8796

[62] Goebel DM, Katz I. Fundamentals of Electric Propulsion. Hoboken, NJ, USA: Ion and Hall Thrusters John Wiley & Sons; 1999

[63] McGuire ML, Hack KJ, Manzella DH, Herman DA. Concept designs for NASA's Solar Electric Propulsion Technology Demonstration Mission. American Institute of Aeronautics and Astronautics; 2014. Available from: https://ntrs.nasa.gov/api/citations/20140017761/downloads/20140017761.pdf

[64] Tarifa MC, Pizzuti L. Theoretical Performance Analysis Of Hybrid Rocket Propellants Aiming At The Design Of a Test Bench And a Propulsive System. 8th European Conference for AeroSpace Sciences. 2019. DOI: 10.13009/EUCASS2019-488

[65] Mane S. In-Situ Resource Utilization for Moon & Mars System. International Journal of All Research Education and Scientific Methods (IJARESM). 2022;**10**(10):618–630. Available from: https://www.ijaresm.com/uploaded_files/document_file/Shreya_ManeeaJn.pdf

[66] Hintze PE, Meier AJ, Shah MG, Devor R. Sabatier system design study for a mars ISRU propellant production plant. International Conference on Environmental Systems. 2018. p. 1–11

[67] Robinot J, Rodat S, Abanades S, et al. Review of in-situ oxygen extraction from lunar regolith with focus on solar thermal and laser vacuum pyrolysis. Acta Astronautica. 2025;**234**:242–259

[68] Acker D, Zamudio-Turcotte K, Modi P, et al. The ZEUS constellation - paving the way to sustainability on the Moon with Solar Power Satellites - IAC-24,C3,4,11,x87582. Acta Astronautica. 2025;**235**:320–338

[69] Buitrago-Leiva JN, Camps A, Moncada NA. Considerations for Eco-LeanSat satellite manufacturing and recycling. Sustainability. 2024;**16**:4933. DOI: 10.3390/su16124933

[70] Rimani J, Paissoni CA, Viola N, et al Multidisciplinary mission and system design tool for a reusable electric propulsion space tug. Acta Astronautica. 2020;**175**:387–395

[71] Space Nuclear Propulsion for Human Mars Exploration. Washington, D.C: National Academies Press; 2021

[72] Krishnan S. Hydrogen Peroxide / Kerosene, Liquid-Oxygen / Kerosene, and Liquid-Oxygen / Liquid Methane

for Upper Stage Propulsion (Technical report). Skudai, Malaysia: Universiti Teknologi Malaysia; 2010. Available from: https://propulsion-skrishnan.com/pdf/Three%20Bipropellants%20for%20Upper%20Stage%20Propulsion.pdf

[73] Clark S. NASA satellite set to conclude successful green propellant demo mission. Spaceflight Now. 2019

[74] Frąckiewicz M. Rocketing into the new space race: Inside the global boom of private spaceflight and payload companies. TS2 China, Internet, Polit. Priv. Ind. Sp. Explor. Technol. Technol. Technol. Trends. 2024

Chapter 3

Space Weather and Civil Aviation

Judith de Patoul, Tobias G.W. Verhulst and Yana Maneva

Abstract

Space weather, including solar flares, coronal mass ejections, solar energetic particles, and solar radio bursts, affects aviation by disrupting high-frequency radio communications (HF COM), degrading Global Navigation Satellite Systems (GNSS), and occasionally increasing radiation levels at cruising altitudes. This chapter reviews the physical basis of space weather, its relevance for aviation, and the International Civil Aviation Organization (ICAO) framework for providing operational space weather information to pilots and air traffic controllers. It also analyses ICAO advisories issued from November 2019 to September 2025 to identify when and where aviation impacts arise. Of 2,350 operational advisories (tests and exercises advisories being excluded), 1,472 concern GNSS and 872 HF COM. The results show that advisory numbers increased with Solar Cycle 25, with marked peaks in May 2024, October 2024, and April 2025. Month-of-year patterns indicate an equinox effect consistent with more frequent geomagnetic disturbances. Geographically, GNSS impacts cluster over the equatorial South Atlantic sector, including parts of South America, whereas HF COM advisories concentrate in the auroral oval and polar cap regions, with a Northern Hemisphere excess. Finally, differences between the four global Space Weather Centres mainly reflect variations in sensors, models, and operational practices. As a possible future service improvement, recommendations are proposed on harmonised trigger-to-action guidance, strengthened high-latitude and airborne observations, real-time integration of radiation dose and GNSS integrity data, and standardised operational logging.

Keywords: space weather, aviation, GNSS, HF communication, ionospheric disturbances, solar flares, coronal mass ejections, geomagnetic storms, solar activity

1. Introduction

1.1 Definition, physical background, and aviation relevance

'*Space weather is the physical and phenomenological state of natural space environments. The associated discipline aims, through observation, monitoring, analysis, and modelling, at understanding and predicting the state of the Sun, the interplanetary and planetary environments, and the solar and non-solar-driven perturbations that affect them, and also at forecasting and nowcasting the potential impacts on biological and technological systems*'. This working definition was elaborated within the European

Cooperation in Science and Technology (COST) Action 724 community, which codified multidisciplinary requirements that link heliophysics, geospace, and impact assessment across user sectors [1, 2].

A wide range of space weather effects is recognised across the global economy. Evidence syntheses for the United States identify structured impact pathways in four priority sectors, namely electric power grids, the satellite industry, Global Navigation Satellite Systems (GNSS) users, and the aviation sector, with simultaneous emphasis on frequent moderate disturbances and rarer extremes with escalating costs and losses [3]. A National Academies workshop report documents potential macroeconomic consequences and cross-sector interdependencies that can amplify impacts when power, communications, and navigation are simultaneously perturbed [4]. Engineering assessments by a national academy describe impacts on engineered systems, including radar, high-frequency (HF) radio communications (HF COM), satellite links, GNSS, and increased solar radiation, with corresponding mitigation principles for service providers and operators, aviation included [5]. Government risk guidance lists space weather as a civil contingency risk with expected consequences for essential services such as air travel, communications, and energy, and summarises characteristic timescales and effects associated with flares, solar energetic particle events, and geomagnetic storms [6].

The primary driver for space weather is the variable Sun [7]. Solar flares, producing mostly hard X-rays but also extreme ultraviolet and/or soft X-ray flux, coronal mass ejections (CMEs), solar energetic particles (SEP), solar radio bursts, and high-speed solar-wind streams (HSS) perturb the heliosphere and geospace [8–12].

For civil aviation, coupling to geospace produces ionospheric electron density structuring across scales that affects HF propagation, total electron content (TEC), the stability and quality of GNSS signals, and, on rare occasions, initiates energetic particle impacts at various flight altitudes in the polar and auroral regions as a function of spectrum and geomagnetic rigidity cutoff. Operational exposure is also strongly structured by geomagnetic latitude. Four broad zones are typically distinguished: the polar cap (≳75°), the auroral oval (~60–75°), mid-latitudes (~40–60°), and low latitudes (≤ ± 30°). Disturbances often localise within one or two of these zones, with additional modulation by local time, season, and storm phase. This geographic context helps map aviation exposure on polar, mid-latitude, and equatorial routes and frames the pathways summarised below. These mechanisms underpin four principal pathways. First, solar flares (on the sunlit hemisphere) and particle precipitation – due to SEP events in the polar cap or from the reconnection region in the magnetotail during geomagnetic storms – produce enhanced ionisation in the lower ionosphere (D-region), resulting in absorption of HF radio signals affecting different regions (see **Table 1**).

Second, post-storm depletion of the electron density in the F-region results in a depression of the maximum usable frequency for HF communication [11, 13–15]. Third, small-scale structures in the ionosphere produced during geomagnetic storms, as well as equatorial plasma bubbles (EPBs), generate GNSS scintillation, cycle slips, and group-delay errors that degrade accuracy, integrity, and availability for navigation and timing [16, 17]. Fourth, SEP events elevate effective dose rates at flight altitudes, with access controlled by geomagnetic cutoff and interplanetary

Space weather phenomenon	Source of D-layer ionisation	Region affected	Typical time-scale (duration)
Solar flare	Hard X-rays	Dayside mid and low latitudes	Tens of minutes
SEP event	Energetic particles	Polar cap (from about 80° latitude)	Hours
Geomagnetic storm	Energetic particles	Auroral oval (about 60° to 80° latitude)	Hours to days

Table 1.
Sources of HF radio absorption due to increased ionisation in the D-region (80 to 100 km altitude) of the ionosphere ('Typical time-scale (duration)' is the latency from solar origin to geospace/ionospheric impact; the subsequent absorption duration depends on event strength, local time, and storm phase).

field configuration, which has implications for crew dose management and avionics single-event effects [12, 18]. These pathways are corroborated by aviation-salient episodes that include cross-polar rerouting on 23 January 2012 owing to HF uncertainties, significant degradation of European GNSS augmentation during September 2017, and a strong 4 November 2015 solar radio burst near 1 GHz that disturbed secondary surveillance radars (SSR) at several European airports [19–21].

Space weather is increasingly recognised as a sustainability-relevant resilience stressor for aviation because it can simultaneously degrade communications, navigation, and radiation margins, forcing operational mitigations that propagate into efficiency and emissions penalties. Evidence from past events shows direct continuity impacts: during the October to November 2003 storms, the FAA's WAAS SBAS capability was disabled for about 30 hours, illustrating how ionospheric disturbance can temporarily remove precision navigation services over wide areas, with knock-on effects for capacity and operational stability [4]. Rerouting efficiency is similarly affected: in January 2005, 26 United Airlines flights were diverted to avoid HF radio blackouts during polar cap absorption conditions, demonstrating how risk controls can shift aircraft away from optimum tracks when high-latitude communications become unreliable [4]. These mitigations translate into tangible cost and fuel exposure; for the United States airline sector, HF-driven rerouting and cancellations are estimated to produce event-scale costs of roughly $0.4 M to $5 M for moderate conditions and $1 M to $30 M for more extreme scenarios, reflecting longer tracks, added fuel burn, delays, and network disruption [3]. Beyond costs, these cases underscore passenger safety and confidence: the primary purpose of disabling augmentation or rerouting away from polar routes is to preserve integrity, reliable communications, and acceptable operational risk, but repeated irregular operations can undermine predictability and trust [4]. From a long-term sustainability perspective, robust space weather practices therefore function as 'efficiency-preserving' resilience measures: improved monitoring, harmonised trigger-to-action procedures, and systematic logging of operational outcomes can reduce both under-warning (safety risk) and over-warning (unnecessary inefficiency), enabling better calibrated decisions that minimise avoidable diversions and associated emissions [22].

1.2 International Civil Aviation Organization framework and global space weather centres

The International Civil Aviation Organization (ICAO) incorporated space weather into its Standards and Recommended Practices (SARPs) for meteorological (MET) service in 2018, with operational implementation on 7 November 2019. ICAO space weather advisories are issued when phenomena are expected to affect HF COM and satellite communications (SATCOM), GNSS-based navigation and surveillance, or to increase radiation exposure at various flight levels. ICAO Annex 3 and the *Manual on Space Weather Information in Support of International Air Navigation* (Doc 10100, see **Figure 1**) define two intensity levels, moderate and severe, the monitored parameters, a standard geographic grid of 30° latitude by 15° longitude, and forecast rows at +6, +12, +18, and +24 hours [23, 24]. The parameter set spans soft X-ray flux for flare classification, ionospheric diagnostics for GNSS impact assessment including The TEC and scintillation indices, S_4 and σ_ϕ, polar-cap and auroral absorption measures for HF COM, and effective dose-rate estimates at flight levels constrained by ground and space observations [24].

Within airline and Air Navigation Service Provider (ANSP) operations, measurable indicators are mapped to actions. For communications, pre-tactical HF planning and alternates are tied to short-wave fadeout (SWF) and absorption risk; for navigation and timing, flight-watch tracks TEC gradients, the Rate Of TEC Index (ROTI), and scintillation indices $S_4$and σ_ϕ together with Satellite-Based Augmentation System (SBAS) and Ground-Based Augmentation System (GBAS) integrity flags and minima; for radiation, dose-aware dispatch relies on modelled ambient dose equivalent rate with rapid assessment during SEP onsets and rare ground-level enhancements (GLEs), and maintenance logs correlate avionics single-event effects with particle environments as described in the ICAO Doc 10100, *Manual on Space Weather Information in Support of International Air Navigation* (see **Figure 1**). At the governance layer, the World Meteorological Organization (WMO) Expert Team on Space Weather (ET-SWx) advances standardisation, verification and interoperability so space weather advisories can be treated and used like MET products.

Impact Area	Parameter (Unit)	Moderate	Severe
GNSS	Amplitude scintillation S4 (dimensionless)	0.5	0.8
	Phase scintillation σ_ϕ (radians)	0.4	0.7
	Vertical TEC (TEC Unit)	125	175
Radiation	Effective dose (μSievert/hour)	30	80
HF	Auroral absorption (Kp)	8	9
	PCA (dB from 30 MHz riometer data)	2	5
	Solar X-ray (W/m^2) (0.1–0.8 nm)	10^{-4}	10^{-3}
	MUF (%)	30	50

(a) (b)

Figure 1.
ICAO doc 10100, Manual on Space Weather Information in Support of International Air Navigation.

Four global Space Weather Centres (SWXCs) meet these specifications and operate in a fixed two-week rotation with layered redundancy: the Pan-European Consortium for Aviation Space Weather User Services (PECASUS), the Australia–Canada–France–Japan consortium (ACFJ), the National Oceanic and Atmospheric Administration (NOAA) Space Weather Prediction Center (SWPC), and the China–Russia Consortium (CRC); see **Figure 2**. Roles cycle every second Tuesday at 08:00 UTC via a formal handover: the On-Duty Centre (ODC) transfers to Maintenance and Observation Centre (MOC), the Primary Backup Centre (PBC) becomes the new ODC, the Secondary Backup Centre (SBC) becomes the PBC, and the previous MOC becomes the SBC. The ODC is responsible for 24/7 monitoring and issuance of ICAO space weather advisories on the Aeronautical Fixed Service; the PBC maintains ful situational awareness and parallel operations to assume ODC duties immediately if required; the SBC must be ready to assume the PBC (or ODC) role within approximately 8 hours; and the MOC maintains awareness and can step up to the SBC within roughly 36 hours. Routine handover briefings, daily situation summaries, and contingency procedures ensure continuity, consistency, and traceability of advisories across centres. As of August 2025, the four global centres are being further supported by the ICAO regional warning centre in Africa, the South African National Space Agency (SANSA), which provides supplementary data to support advisory issuance.

Figure 3 shows an example of an ICAO space weather advisory issued by the PECASUS Centre on the 5th of December 2021 for an ongoing ionospheric disturbance.

1.3 Purpose and contributions of this study

This chapter provides a global synthesis of civil aviation during Solar Cycle 25 (SC25) and a data-driven assessment of ICAO space weather advisories. Section 1 defines space weather, outlines the heliophysics most relevant to aviation, and sets the regulatory context through ICAO Annex 3 and Doc 10100 so that subsequent analyses use the same parameters, thresholds, and spatial grid. Section 2 links physics

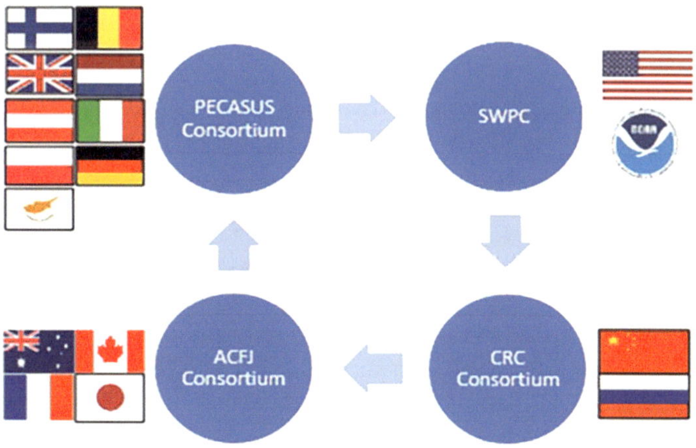

Figure 2.
Global SWXC rotation and composition of the PECASUS, ACFJ, SWPC, and CRC consortia.

```
0000060501
FNXX01 EFKL 051858
SWX ADVISORY
DTG: 20211205/1858Z
SWXC: PECASUS
ADVISORY NR: 2021/49
SWX EFFECT: GNSS MOD
OBS SWX: 05/1821Z HNH E000 - E045
FCST SWX +6 HR: 06/0100Z NOT AVBL
FCST SWX +12 HR: 06/0700Z NOT AVBL
FCST SWX +18 HR: 06/1300Z NOT AVBL
FCST SWX +24 HR: 06/1900Z NOT AVBL
RMK: SPACE WEATHER EVENT (IONOSPHERIC
DISTURBANCE) IN PROGRESS. IMPACT ON GNSS PERFORMANCE
POSSIBLY LEADING TO LOSS OF GNSS SIGNALS AND/OR DEGRADATION
OF TIMING AND POSITIONING PERFORMANCE.
NXT ADVISORY: WILL BE ISSUED BY 20211206/0045Z=
```

Figure 3.
Example ICAO space weather advisory in the standard structure (here GNSS 2021/49).

to operations: it establishes the solar-cycle backdrop, then traces how flares drive short-wave absorption on HF paths, how SEPs raise effective dose rates and induce single-event effects, and how geomagnetic storms restructure the ionosphere to degrade GNSS; the section closes by explaining the typical minutes-to-days sequencing that causes overlapping risks. Section 3 turns to evidence: using the complete ICAO advisory record from November 2019 to September 2025, it quantifies volumes by domain and centre, isolates storm-driven surges around May 2024, October 2024, and April 2025, documents equinox-season maxima, and maps the geographic footprint, including the equatorial South America–Atlantic GNSS hotspot and high-latitude HF belts, while analysing centre-to-centre variability arising from sensors, models, and workflows. This section also distils lessons and limitations, highlighting gaps in polar sensing, real-time radiation monitoring, validation against operational outcomes, and consistency of public communication. Section 4 integrates these findings into concrete recommendations on harmonised trigger-to-action guidance, strengthened observing and nowcasting, standardised logging and data sharing, and the sustainment of global advisory services.

2. Solar cycle context, drivers, indicators, and aviation impacts

2.1 Solar cycle

Solar activity varies quasi-periodically with an 11-year cycle; the sunspot number is a robust proxy for the emergence rate of strong magnetic fields and correlates with flare, CME, and SEP occurrence [25]. In SC25, the daily and monthly international sunspot numbers rose rapidly from the 2019 minimum, with the monthly smoothed curve surpassing the SC24 peak and reaching about 150–165 in late 2024 to early 2025 (see **Figure 4**). Repeated daily bursts above 200 indicate intervals of numerous, magnetically complex active regions. Elevated sunspot levels statistically imply higher rates of eruptive drivers, increasing the probability that aviation-relevant hazards overlap within the same operating week [7, 22].

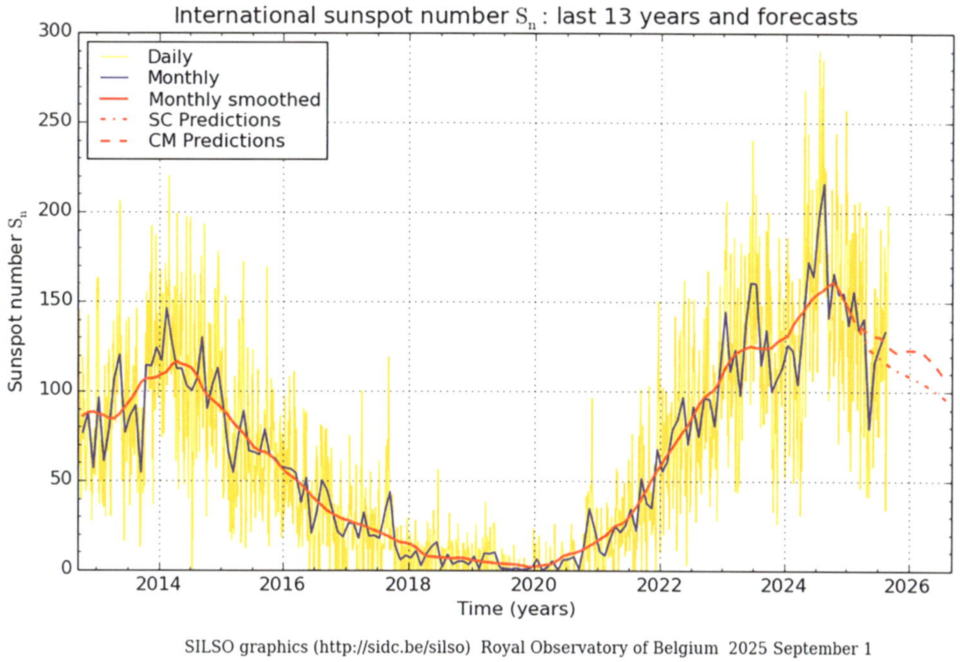

Figure 4.
International sunspot number S_n for the last 13 years with daily (yellow), monthly (blue) and monthly smoothed (red) curves, plus consensus forecasts for the SC25 decline.

The Sunspot Index and Long-term Solar Observations (SILSO) record shows that SC25 activity is organised into multi-day clusters as large active regions transit the disk or multiple regions erupt in succession. For operations, this produces a characteristic sequence: day-one X-ray flare–driven short-wave absorption on sunlit HF paths, hour-scale SEP onsets that raise dose rates at high latitude and altitude and can induce polar cap HF absorption, and day-scale CME or HSS arrivals that trigger geomagnetic storms, which can result in auroral absorption and degrade GNSS performance; the recovery phase of a storm can include post-storm maximum usable frequency (MUF) depressions lasting for several days. These staggered timescales – minutes, hours, and days – explain why impacts on communications, navigation, and radiation exposure often co-occur during active periods.

2.2 HF communication (3–30 MHz)

HF communications impact includes (1) flare-driven SWF on sunlit paths, (2) absorption due to particle precipitation in the auroral oval and polar cap, and (3) post-storm F-region depressions that reduce MUF [13–15, 44–46].

Figure 5 illustrates normal ionospheric reflection, flare-related absorption on the dayside, and high-latitude cases where particle precipitation prevents reflection, with recovery often accompanied by MUF depressions. Solar flares raise the hard X-ray flux and enhance D-region ionisation on the sunlit hemisphere, producing SWF and in the worst cases complete radio blackouts on aviation HF links that remain vital for oceanic operations. Empirical characterisations show that moderate and severe absorption events occur frequently on solar-cycle timescales, last from

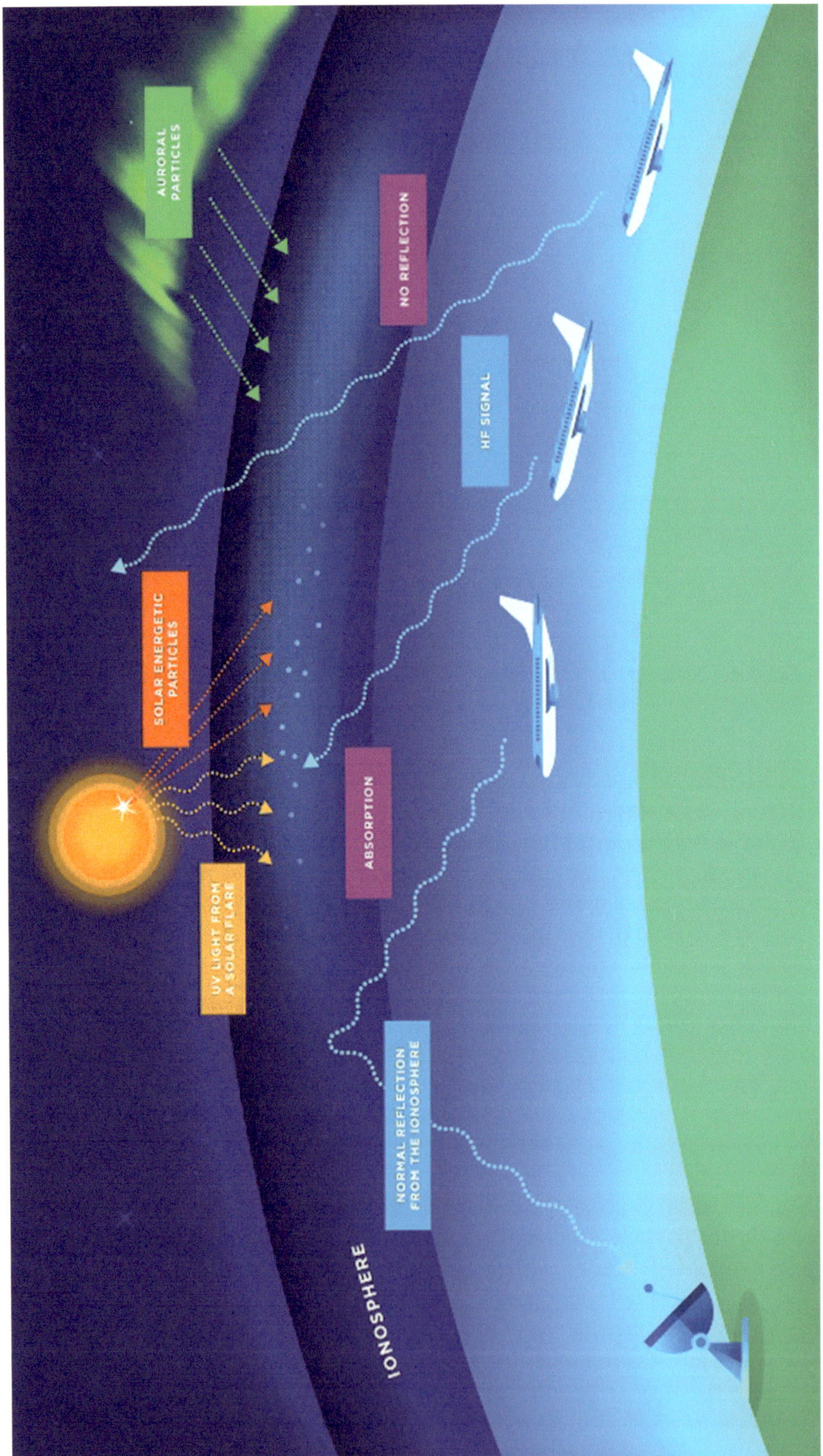

Figure 5.
HF communications under quiet and disturbed ionospheric conditions.

tens of minutes to several hours, and express most strongly at low latitudes and over long over-water sectors. The principal indicators are Geostationary Operational Environmental Satellites (GOES) X-ray flux class and D-region absorption maps; practical cues are observed/forecast SWF probabilities and regional absorption (in dB) in the 10–15 MHz aviation HF bands.

Auroral electrons during substorms and SEPs during proton events increase D-region ionisation at high latitudes, producing auroral absorption and polar-cap absorption. These events are often more severe and longer-lived than flare SWF, with durations ranging from hours to days, and they can be followed by negative F-region storms that depress MUF for 12–72 hours and constrain HF frequency selection.

For all HF impacts, mitigation relies on daylight-path and latitude awareness, pre-tactical frequency planning with alternates, route-dependent use of SATCOM or very high-frequency (VHF) radio (30–300 MHz) relays where feasible, and monitoring of centre products that map absorption and MUF constraints relevant to aviation operations.

2.3 Satellite communication

SATCOM disruptions during extreme bursts can accompany active periods: intense solar radio bursts raise the noise floor near navigation/communication bands and, together with high-latitude structure, can reduce link margins. Such L-band-relevant bursts are rare but consequential; monitoring dynamic spectral flux and link-margin telemetry supports alternate paths, store-and-forward, and priority handling of critical messages [26].

On 4 November 2015, SSR displays at several sites in southern Sweden showed Sun-aligned false plots and intermittent track loss, prompting LFV to curtail movements for safety. Two disturbance windows were reported: 14:19–14:34 UTC and 14:47–14:50 UTC, with the first interval being the most severe. Similar Sun-aligned artefacts were documented in Belgium, while Norway reported milder effects. In all cases, the anomalies occurred when antennas pointed toward the Sun, consistent with front-end overloading near the 1030/1090 MHz SSR channels [21].

At the same time, solar observations show an exceptionally strong, coherent solar radio burst peaking close to 10^5 solar-flux units in a relatively narrow band around 1 GHz, beginning about 30 minutes after the soft X-ray peak of the associated flare. The burst's fine spectral structure (pulsations, fibre, and zebra patterns) and timing implicate coherent coronal processes rather than synchrotron emission. Geomagnetic conditions were only minor–to–moderate (Kp <6), providing no competing magnetospheric explanation for the radar effects.

From a susceptibility standpoint, solar radio bursts at L-band frequency and adjacent ultra-high frequency (UHF) are a recognised threat to aviation systems when flux densities approach 10^5 sfu, with documented cases of GNSS and communications degradation, and evidence that instrument saturation can lead to under-reporting of extremes in historical patrol data [26]. The Swedish episode illustrates a surveillance-specific hazard pathway independent of D-region absorption or GNSS scintillation: a narrowband, high-flux solar radio burst directly elevates the noise floor in the SSR passband and overloads receiver/processing chains, creating false plots and delayed or lost tracks.

Practical mitigations derived from these analyses include immediate cross-checks with primary radar when Sun-aligned artefacts appear, temporary restrictions in affected azimuth sectors, and use of calibrated whole-Sun UHF/L-band flux monitoring to confirm a solar origin and support timely advisory and flow-management decisions.

We should note that SATCOM and solar radio burst advisories are currently not part of the official space weather service to aviation but have been considered for possible future service development.

2.4 Radiation effect and single event effect in avionics

SEP events inject high-energy protons and heavier ions that can elevate effective dose rates above the galactic cosmic-ray baseline, most strongly on polar routes and at higher cruising levels (see **Figure 6**). Observations and modelling document operationally significant enhancements during major SEP episodes and emphasise rapid assessment for rare GLEs, together with exposure management for crew classified as radiation workers [27, 28, 41, 43]. Operational indicators are integral proton fluxes >10, >50, and >100 MeV, and neutron-monitor responses for GLEs; decision cues are modelled effective dose rates ($H^*(10)$) at planned altitudes/latitudes and onset–decay trends.

Mitigations include latitude and altitude optimisation, dose-aware dispatch, and enhanced avionics fault monitoring during elevated flux intervals. The same particle environment produces single-event effects in microelectronics, motivating resilient design, error-correcting code (ECC)/watchdog strategies, and systematic, time-stamped anomaly logging.

2.5 GNSS degradations

GNSS performance is degraded when geoeffective CMEs and HSS restructure the ionosphere (see **Figure 7**). Large-scale TEC gradients and small-scale irregularities introduce group-delay error and carrier-phase scintillation, causing loss of lock and increased cycle slips. This reduces accuracy, integrity, and availability for required navigation performance and for augmented approaches using SBAS or GBAS, and it can affect timing-dependent surveillance such as Automatic Dependent Surveillance–Broadcast (ADS-B). At low latitudes, post-sunset EPBs can drive strong phase scintillation and loss of lock [29]. Importantly, EPBs are not necessarily storm-driven; they often occur under quiet-time pre-reversal enhancement conditions, and strong storms can sometimes suppress EPB formation.

Mitigation options include using multi-constellation, multi-frequency GNSS (GPS, Galileo, BeiDou, GLONASS on L1/E1 and L5/E5), enabling receiver autonomous integrity monitoring (RAIM)/advanced RAIM (ARAIM), fault detection and exclusion (FDE), and receiver reversion logic; reverting to higher minima (for example, LNAV or Baro-VNAV) or to conventional navaids (VOR/DME/ILS) where available; coordinating with ATC for vectoring and spacing; and employing surveillance fallbacks such as secondary radar or ADS-C via SATCOM when ADS-B is impaired. Special attention is given to equatorial post-sunset scintillation windows and to polar-cap dynamics. During recovery, negative F-region storms can persist and lower the MUF for HF COM, so GNSS mitigations are aligned with HF communications planning.

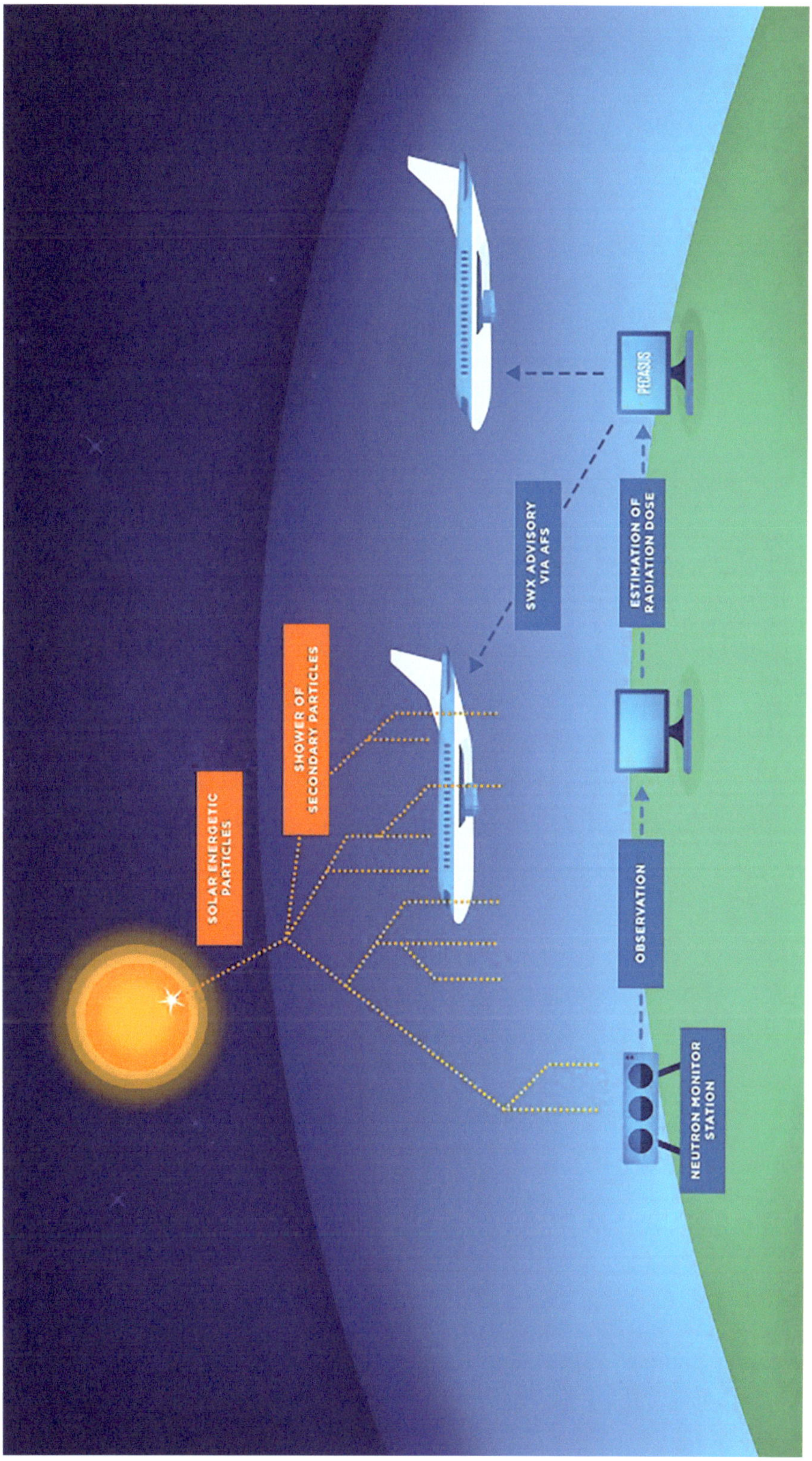

Figure 6.
SEP radiation environment at flight levels and advisory workflow.

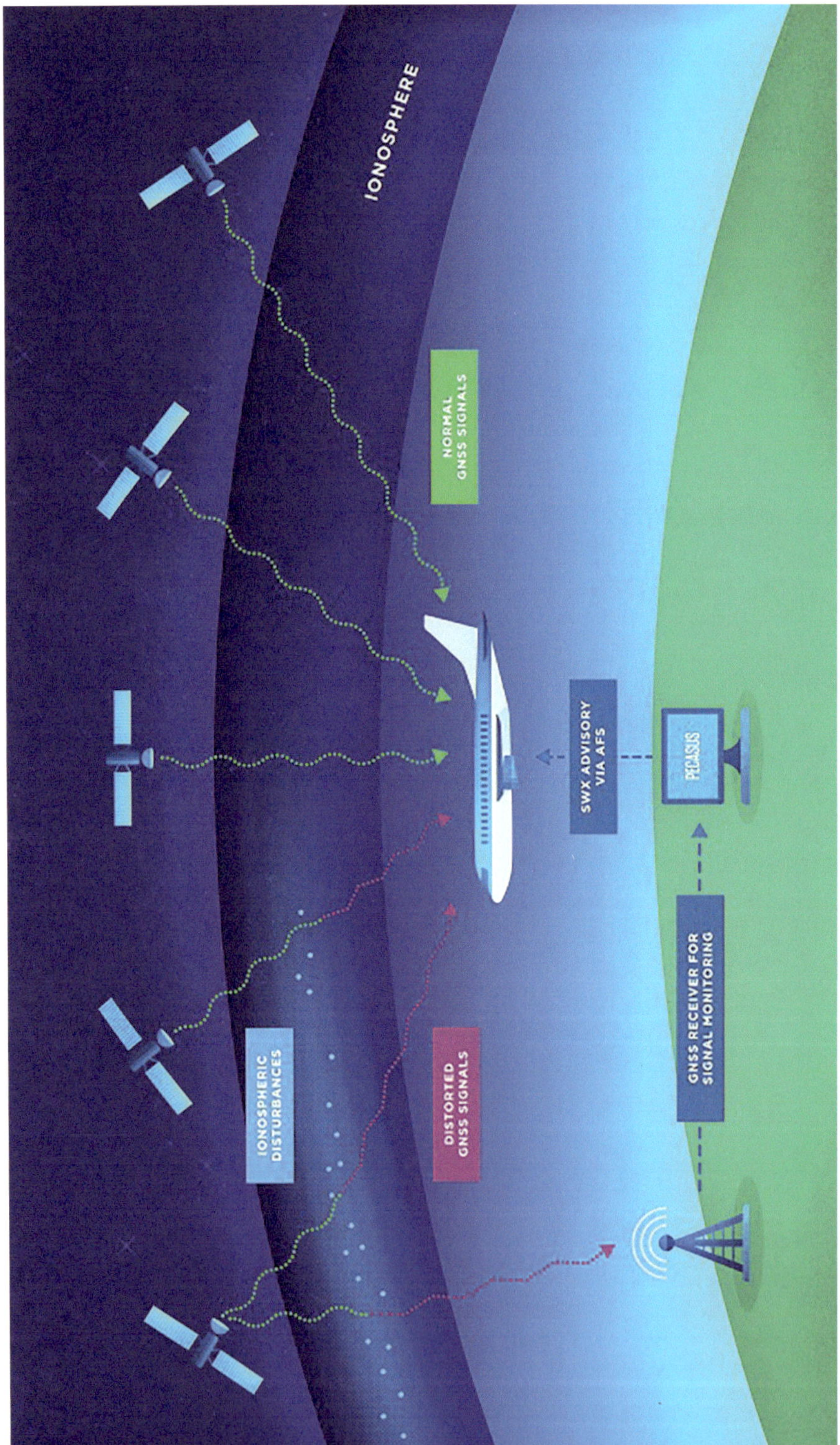

Figure 7.
Ionospheric disturbances and the pathway to GNSS degradations.

2.6 Multi-driver sequencing with an operational implications example

Active periods commonly unfold as a sequence: an initial flare produces immediate HF COM impacts on sunlit paths; hours later, a significant eruption may yield an SEP onset that elevates dose rates; several days later, CME arrival triggers geomagnetic storms that degrade GNSS [47], with HF COM recovery further limited by post-storm MUF depressions. This minutes-to-days staging explains why the risk to communications, navigation, and radiation exposure often overlaps for a week or more. It underpins the value of impact-based advisories and airline/ANSP procedures that bind measurable indicators to harmonised trigger-to-action responses [7, 15, 20].

To illustrate advisory sequencing and operational implications, we cite the May 2024 active window (6 to 15 May 2024), during which the PECASUS SWXC issued a dense HF COM advisory sequence. During this period, 10 flare-driven HF COM SWF advisories between 6 and 13 May were issued, followed by storm-time HF impacts, including polar cap absorption (PCA) on 10 to 13 May, auroral absorption (AA) on 10 to 11 May, and a post-storm depression (PSD) phase lasting from 10 to 15 May [30]. This ordering matches the characteristic staging: rapid sunlit HF COM degradation during flares, then high-latitude absorption during the storm main phase, and finally a multi-day recovery period where MUF depression can keep HF COM usability reduced even after peak geomagnetic forcing. The PSD tail, therefore, increases the probability of co-occurrence with GNSS degradation windows, extending the period in which operators may face cross-domain constraints.

Such overlaps can amplify downstream constraints in structured airspace systems. For North Atlantic operations, mitigation actions aimed at retaining HF COM and GNSS integrity can include southward displacement of the North Atlantic Organised Track System (OTS), which reduces available airspace volume, increases sector loading at lower latitudes, and can propagate network-wide delay through flow and capacity management [31]. From an operator's perspective, an HF-heavy sequence lasting approximately nine days (6 to 15 May) compels earlier and more conservative communications planning on oceanic and polar segments, including frequency management and timely activation of alternates (SATCOM and VHF where available), and may motivate tactical timing or track choices with fuel and schedule penalties. Linking advisory series to operational outcomes (delay, diversions, and communications performance) supports calibration of trigger-to-action thresholds and procedures for multi-hazard periods.

3. Outcome of recent years, notable events, and lessons learned

In this section, we analyse the complete operational ICAO Space Weather Advisory (SWX ADVISORY) record disseminated from 7 November 2019 to 30 September 2025, excluding TEST and EXER messages. The dataset is global in intent but encoded on the ICAO standard grid (30° latitude × 15° longitude) and in the ICAO advisory shorthand longitudinal sectors (i.e., HNH/HSH, MNH/MSH, EQN/EQS, and DAYLIGHT SIDE), which imposes a deliberate spatial coarseness. The ICAO system reports impact-based domains (HF COM, GNSS, and RADIATION) and only two severity levels, MOD (moderate) and SEV (severe);

therefore, the advisory record is not a physical-event catalogue but a thresholded operational representation of expected impacts.

Advisories were retrieved programmatically through the ESA Space Weather Service Network portal product S.790a, 'ICAO Space Weather Advisory browser' (https://swe.ssa.esa.int/web/guest/sidc-S790a-federated), using the federated Web API (mandatory parameters: pc = S790 and psc = a). Queries with the component = archive and ISO 8601 UTC start and end timestamps (dts_start, dts_end) for the retrospective record were used; status = no_test was applied to exclude TEST and EXER entries at the source, and responses were ingested as JSON.

Each advisory was parsed to extract the issuance centre (SWXC), timestamp (OBS SWX), phenomenon type and severity (SWX EFFECT), and forecast rows (FCST SWX +6, +12, +18, +24 HR); see **Figure 3**. We define an advisory series as an opening advisory plus its updates and closing, using the Number/Replaces 'NR RPLC' logic, which we use as a practical proxy for a distinct operational episode, noting that unique physical start and end times cannot be recovered from the advisory record alone.

3.1 Global results from November 2019 to September 2025

From November 2019 to September 2025, excluding TEST (test) and EXER (exercise), 867 advisory series and 2350 operational space weather advisories were disseminated, comprising 1,592 moderate and 758 severe. By domain, GNSS accounted for 1,472 (62.7%), HF COM 872 (37.1%), and radiation 6 (0.26%). An advisory series – opening, updates, and closing – serves as a practical proxy for a distinct observed event, acknowledging that exact physical start and end times cannot be uniquely recovered; by series, GNSS comprised 587 (67.7%), HF COM 278 (32.1%), and radiation 2 (0.23%). Within a series, updates are issued every three to six hours and never later than six hours after the previous advisory. By issuing centre, PECASUS produced 1,111 (47.3%), ACFJ 998 (42.5%), SWPC 133 (5.7%), and CRC 108 (4.6%). **Figures 8** and **9** summarise these results.

Following post-event analysis, the six radiation advisories were determined to be false alerts arising from model errors that have since been corrected; accordingly, the

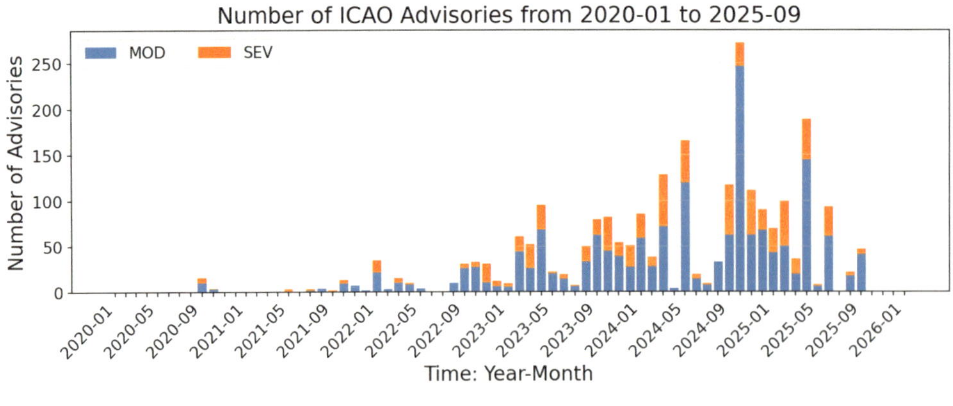

Figure 8.
Advisory count by severity (SEV for severe, orange; MOD for moderate, blue), stacked by month across all centres.

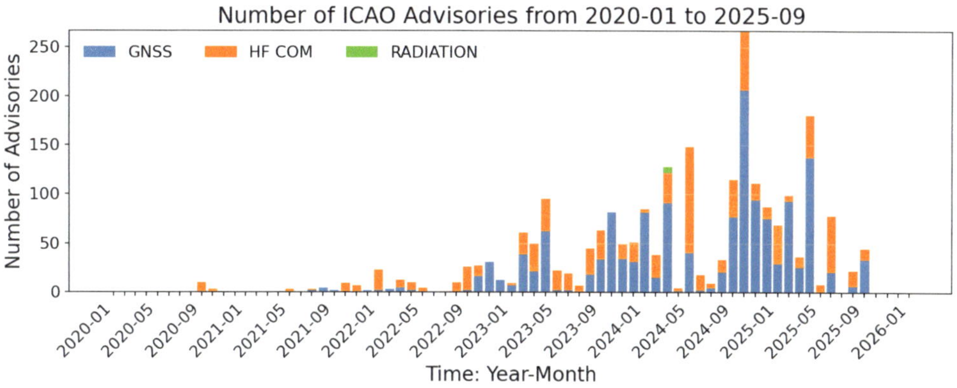

Figure 9.
Advisory count by domain (GNSS in blue, HF COM in orange, RADIATION in green), stacked by month across all centres.

remainder of this chapter focuses primarily on GNSS and HF COM events and advisories. First, from the 2019 minimum to the late-2024/early-2025 maximum of SC25, the SILSO sunspot record shows a strong rise in daily and monthly counts and a high monthly smoothed plateau (~150–165). On this decadal scale, advisory production follows the same envelope: monthly totals for all centres climb through 2023–2025 and peak near the SC25 maximum (see **Figure 10**). This is physically expected because the sunspot number is a robust proxy for the emergence of strong magnetic fields and, statistically, for rates of flares, CMEs, and SEP events – the principal drivers behind HF absorption, GNSS degradations, and radiation episodes [25]. The correspondence is not one-to-one: Issuance depends on threshold exceedance and geographic coverage by ICAO's grid, and on the mixture of drivers in any month, as we will see below. But on the 11-year trend, the co-variation between sunspot activity and advisory volume is clear and consistent with system-level climatology that links cycle phase to disturbance incidence across the Sun–heliosphere–ionosphere chain [7].

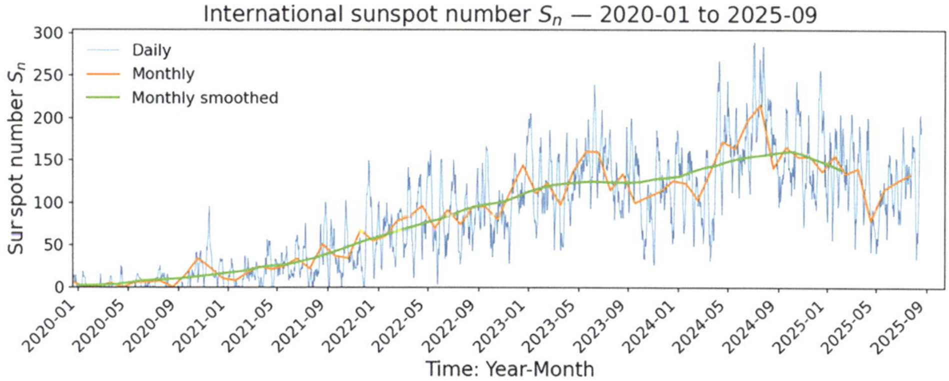

Figure 10.
International sunspot number S_n from January 2020 to September 2025.

Second, from November 2019 to September 2025, the advisory record shows three dominant storm-related surges during SC25. The first occurred on 10–11 May 2024 during an extreme geomagnetic storm (Kp = 9); that month accumulated 148 advisories in total (GNSS 40; HF COM 108), reflecting strong GNSS structuring and extensive auroral/polar-cap absorption with post-storm MUF depression for HF. The second peaked on 10–11 October 2024 with a severe storm (Kp = 8) following a fast CME launched on 8 October. October totals were 266 advisories (GNSS 206; HF COM 60), consistent with storm-time TEC gradients and scintillation driving GNSS integrity/availability degradations alongside HF recovery constraints. The third centred on 16 April 2025, again severe (Kp = 8-) from multiple CMEs erupting from 13 April. April totals were 180 advisories (GNSS 137; HF 43), indicating widespread GNSS impacts with concurrent high-latitude HF constraints. Radiation advisories were absent in these months.

Third, the results highlighted an imbalance in the number of advisories issued by each centre. This is primarily attributable to heterogeneity in (1) observing networks used, (2) modelling stacks, and (3) operational workflows, not to formatting guidance. Each Space Weather Centre (SWXC) ingests a different mix of ground and space data. Densities and latitudinal coverage of ionosondes, riometers, and GNSS receivers vary widely between Europe, North America, Australasia, and Asia; consequently, the same disturbance can be sampled with different fidelity in different regions. Centres that exploit dense regional GNSS networks can generate higher-confidence ROT/ROTI and scintillation (S4, $\sigma\varphi$) fields and therefore issue more GNSS advisories than centres relying mainly on global products. Likewise, centres with strong high-latitude riometer coverage will detect polar cap and auroral absorption more often, feeding HF communications advisories. Moreover, the modelling layers used to translate raw observations to ICAO parameters are not uniform. Examples include different TEC and ROTI algorithms and thresholds for GNSS integrity risk; alternative absorption proxies for HF COM (e.g., riometer-based versus empirical X-ray/solar-zenith formulations); distinct F-region background models for MUF-depression detection; and different radiation nowcasting frameworks and dose maps for SEP/GLE conditions. These choices directly affect whether Annex 3/Doc 10100 thresholds are crossed in a given grid cell and hour, and thus how many advisories are issued. Finally, each Centres implement different sequencing and quality-control logic: for example, how to merge adjacent ICAO grid cells, how long to persist 'MOD/SEV' conditions in the absence of fresh evidence, when to collapse fluctuating indicators into a single update, and how to terminate advisories. PECASUS's end-to-end chain – data ingest, modelling, validation, and analyst oversight – is one documented implementation and illustrates how pipeline design influences issuance frequency [20].

In order to mitigate this aspect of imbalance in the number of advisories, a jointly internal document, *Guidance for the Issuance of Space Weather Advisories* (Guidance for the Issuance of Space Weather Advisories, 2025 [48]), harmonises templates, phrasing, initiation/update/termination logic and 'smoothing' practices across HF COM, GNSS and RADIATION products. It reduces stylistic and timing differences, improving comparability for users, but it cannot remove fundamental disparities arising from the heterogeneous sensors and models each centre is able (or mandated) to use. As long as the underlying data streams and transformation methods differ, advisory volumes will legitimately diverge even under identical ICAO standards.

The centre-to-centre spread in counts is therefore an expected property of the current ecosystem: it reflects where and how the geospace is observed and modelled, rather than inconsistent adherence to ICAO standards. Any attempt to compare volumes as a performance metric must control for input data, modelling choices, and workflow rules; otherwise, it conflates scientific and infrastructural differences with service quality.

3.2 Seasonal pattern of advisories: Equinox enhancement

Figure 11 (HF COM) and **Figure 12** (GNSS) aggregate all advisories from November 2019 to September 2025 by calendar month (independent of year) and stack them by Space Weather Centre (PECASUS, ACFJ, SWPC, CRC). HF COM totals peak in May (143), September (114), June (91), and April (87). GNSS totals peak in October (304), April (202), November (159), and February (149). Contributions are centre-dependent: for GNSS, the October peak is driven chiefly by ACFJ (214/304); November is dominated by PECASUS (131/159); April is led by

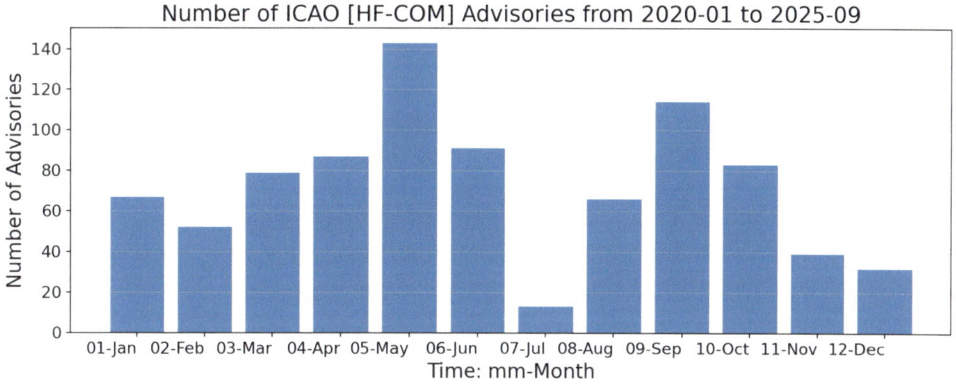

Figure 11.
HF COM, total advisories per month (independent of year), stacked by centre.

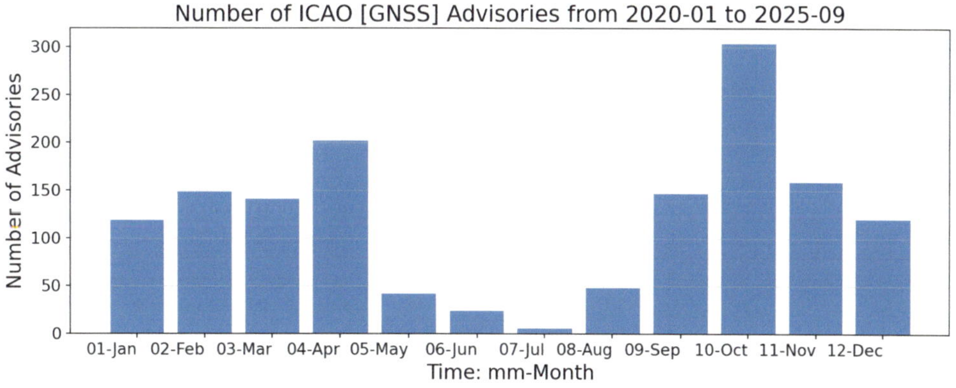

Figure 12.
GNSS total advisories per month (independent of year), stacked by centre.

ACFJ (144/202). For HF COM, the May maximum is shared mainly by ACFJ (76) and PECASUS (37), with SWPC (21) and CRC (9) also contributing; CRC contributes notably in October (32), and SWPC in several spring–autumn months (e.g., February 2024: 17; May 2023: 17; June 2024: 14). This seasonal pattern is consistent with the semiannual (equinox) enhancement of geomagnetic coupling, the Russell–McPherron effect, by which southward interplanetary magnetic field (IMF) projections and coupling functions are statistically larger near March–April and September–October [32, 33]. Stronger coupling elevates the intensity of geomagnetic storms and substorms, driving TEC gradients, scintillation, and loss-of-lock that underpin the GNSS maxima in fall (October and November) and spring (February to April). At low latitudes, equinox seasons also favour post-sunset electrodynamics and EPBs, reinforcing phase and amplitude scintillations, resulting in GNSS advisories. For HF COM, equinox storms often yield post-storm F-region depressions that reduce MUF for 12–72 hours, while flare activity throughout the cycle produces sunlit SWF; together, these produce the HF COM advisories maxima between April and June and in September. In addition to the equinox tendency, the monthly distributions show clear differences between the June and December solstices. Part of this contrast is physical: post-sunset EPB occurrence and growth rates differ between solstices because the magnetic equator is offset from the geographic equator, modifying electrodynamic preconditioning. However, part of the difference may reflect observing-network asymmetries, as data availability and coverage are not strictly balanced between hemispheres in our inputs. Accordingly, solstice contrasts should be interpreted with caution pending a dedicated hemispheric analysis.

Several storm months amplify the monthly bars beyond the seasonal effect, most prominently May 2024, October 2024, and April 2025, so **Figures 11** and **12** reflect both the equinox tendency and the contribution of specific severe events during those months. This is visible despite the year-independent binning.

3.3 Geographic footprint of observed advisories

Figure 13 maps how often ICAO grid cells were reported as affected in observed advisories from November 2019 to September 2025. For each advisory, the ICAO standard grid footprint (30° latitude by 15° longitude cells) was decoded, and each flagged cell was counted once; counts were then summed across all domains and all centres. The global pattern shows a low-latitude belt and high-latitude belts being more affected.

Figure 14 isolates GNSS advisories and shows maxima between about 30° S and 30° N with a pronounced peak over South America and the equatorial Atlantic. This pattern reflects the equatorial ionisation anomaly (EIA) created by the daytime $E \times B$ fountain and field-aligned diffusion, which forms two crests near ± 16° magnetic latitude that can migrate toward ±30° under strong forcing [34, 35]. During major storms, prompt-penetration eastward electric fields acting with storm-time equator-ward neutral winds can produce a 'super-fountain' effect, strongly enhancing the ionisation in the Appleton anomaly [36, 37]. During the 30 October 2003 event, crest-to-trough ratios approached ~10, electron density at the crests increased by >1,000%, crests shifted to ~ ± 30°, and dusk upward drifts in Brazil reached ~1,200 m/s; these observations show that electric fields alone are insufficient without concurrent neutral-wind

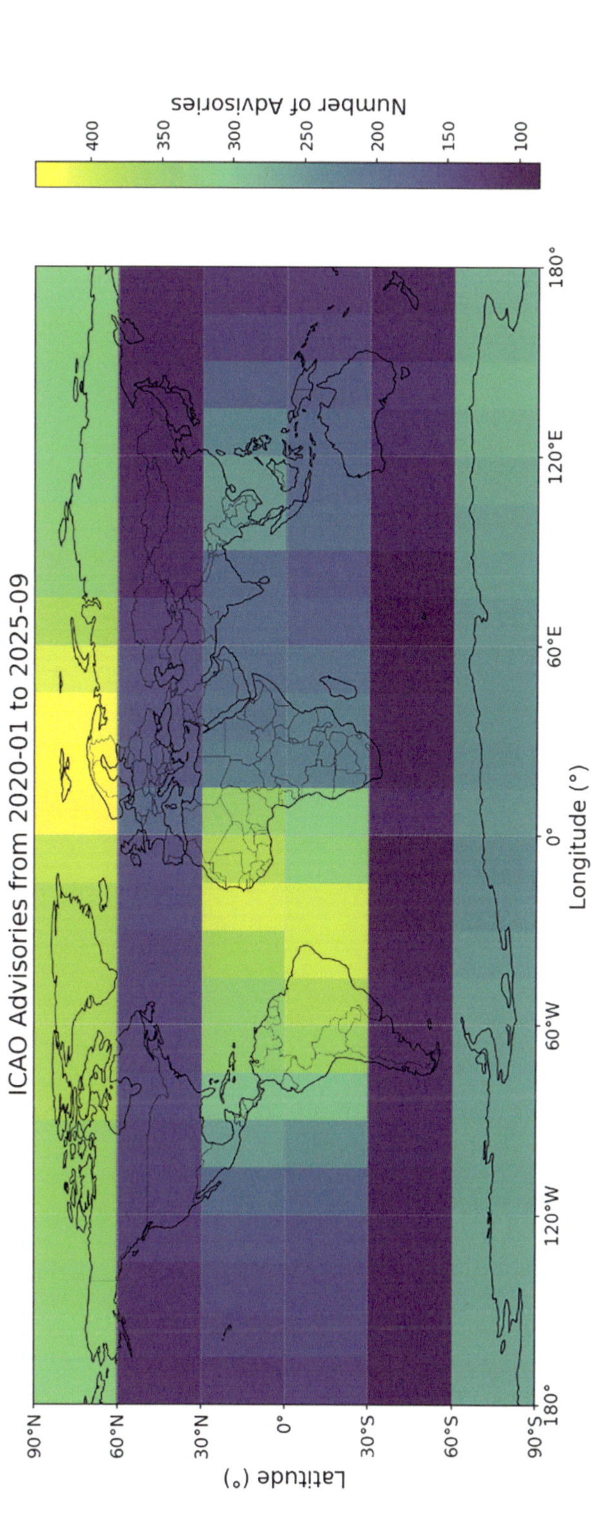

Figure 13.
Aggregate count of ICAO grid cells reported OBS affected by any advisory domain (GNSS, HF COM, RADIATION) across all centres, November 2019–September 2025.

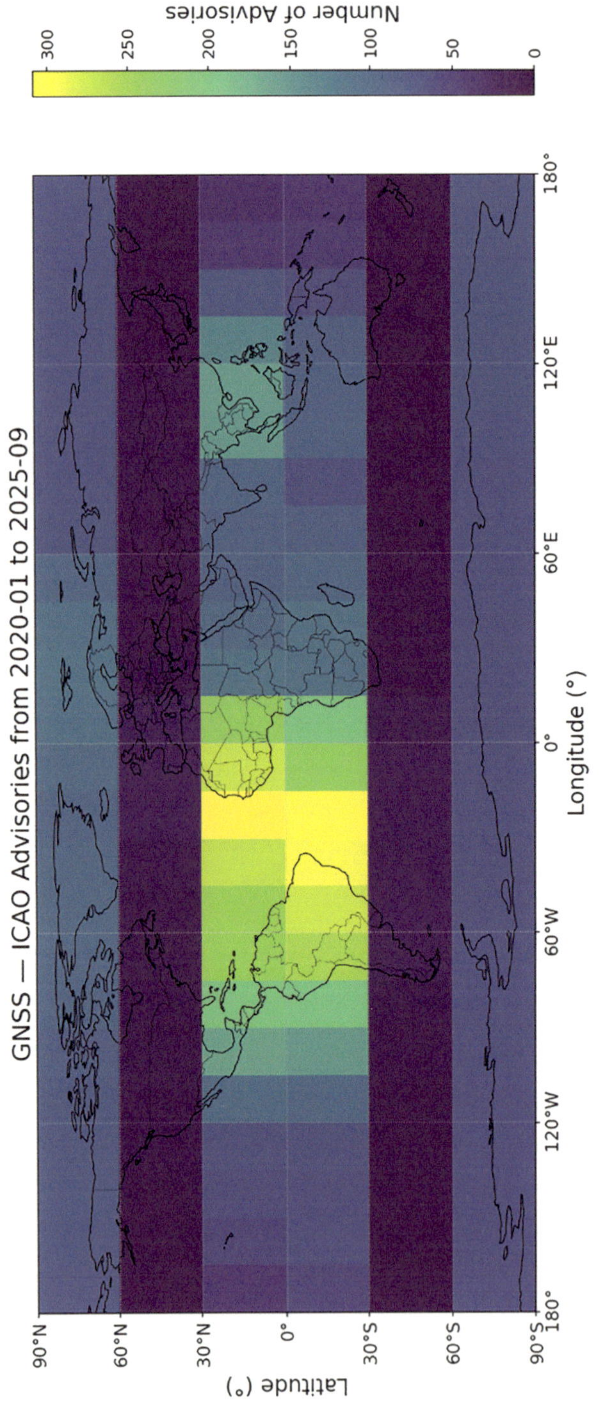

Figure 14.
Count of ICAO grid cells reported OBS affected by GNSS advisories (Nov 2019–Sep 2025).

forcing [35]. After sunset, the pre-reversal enhancement lifts up the bottom-side F layer; once a solar-activity-dependent threshold height of roughly 350–450 km is exceeded and some seed perturbations (often a bottom-side atmospheric gravity wave) exist, generalised Rayleigh–Taylor instabilities at the bottom of the plasma enhancement generate EPBs that move up and map along the anomaly crests. These structures cause strong GNSS phase and amplitude scintillation, cycle slips and loss of lock, with occurrence peaking near the equinoxes and in the American longitude sector, where the peculiar configuration of the geomagnetic field associated with the South-Atlantic anomaly favours the generation of plasma bubbles. This explains the advisory maximum over South America and the adjacent Atlantic in **Figure 14**.

The HF COM-only map (see **Figure 15**) highlights the auroral-oval and polar-cap regions where particle precipitation enhances D-region ionisation, producing HF absorption, and where recovery phases commonly show sustained MUF depressions. A polar dominance of impacts is expected and consistent with reviews of polar-cap absorption, auroral absorption, and post-storm F-region depressions [38, 44–46]. A clear Northern Hemisphere excess appears in our dataset. This aligns with established seasonal and hemispheric asymmetries, including the equinox enhancement of geomagnetic coupling and interhemispheric differences in ionospheric response [32, 33, 39, 49]. Over the past five years, the Northern Hemisphere excess persists, but a dedicated study is required to separate sampling effects, such as observing-network and route density, from geophysical drivers, such as field geometry and winter-hemisphere composition changes.

3.4 Lessons learned: Gaps and challenges

The advisory record highlights structural blind spots in the observing system. High-latitude corridors over the Northern Ocean and North Pacific remain sparsely instrumented for riometers, ionosondes, and dense GNSS networks (in particular, receivers capable of monitoring scintillation), limiting skill for auroral and polar-cap absorption nowcasts and for storm-time TEC gradient and scintillation diagnostics; this under-sampling is most acute exactly where HF COM and GNSS vulnerabilities concentrate. Radiation situational awareness at flight levels is still model-dominant: fleet-wide, low-latency *in situ* dose data are rare, SEP onset detection can lag operational needs, and assimilation frameworks for effective dose rate H*(10) remain uneven across providers.

Centre-to-centre differences in issued volumes persist because input data streams, model stacks, and workflow logic are heterogeneous. Variations in GNSS gradient and scintillation metrics, HF absorption proxies and post-storm MUF-depression detection, plus differing rules for grid-cell painting, persistence and termination, all change whether Annex 3 and Doc 10100 thresholds are crossed in a given hour and location. Recent common guidance harmonises templates and timing but cannot remove disparities rooted in sensor coverage and modelling choices.

The coarse ICAO grid (30° × 15°) introduces spatial smoothing, so advisory footprints can over or under-represent the true impact extent. Moreover, because the grid is defined in geographic coordinates, equal-latitude bands are not comparable across longitude sectors. For instance, the 60°–90° band over the American sector is not equivalent – in area or morphology – to the same band over

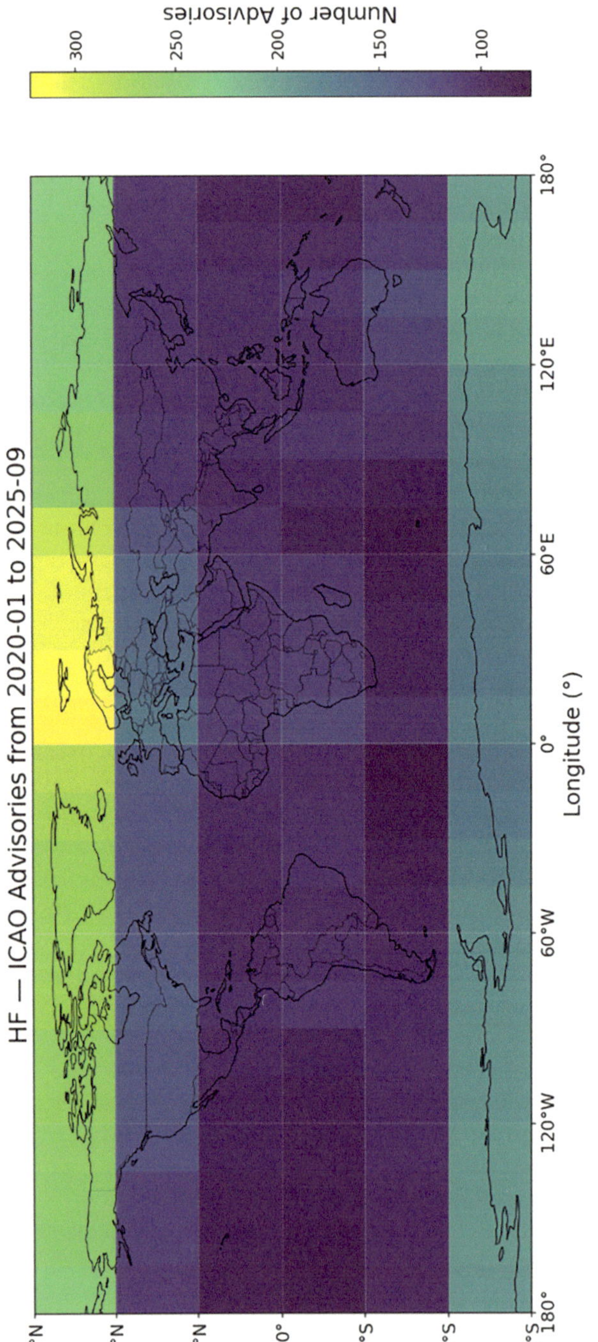

Figure 15.
Count of ICAO grid cells reported OBS affected by HF COM advisories (November 2019–September 2025).

Europe. A finer, preferably geomagnetically referenced or equal-area grid would mitigate this bias. For example, the patterns seen in **Figure 14**, as well as the solar cycle and seasonal dependencies mentioned above, agree well with known variations in occurrences of scintillation in trans-ionospheric radio signals [40]. However, the ICAO grid is so coarse that the appearance of auroral oval and polar cap scintillation – which are produced by different physical phenomena and do not necessarily occur together – cannot be distinguished. Polar-region prominence is physically expected, as seen in **Figure 15**: SEP and auroral precipitation drive D-region absorption that is typically more severe and longer-lived than flare SWF, with post-storm MUF depressions further extending constraints. However, the geographic pattern is likely distorted by data availability: high-latitude corridors remain sparsely instrumented (riometers, ionosondes, dense GNSS/scintillation receivers), and route sampling is uneven across hemispheres and centres. Accordingly, we must treat the polar maxima as conservative, network-limited estimates until expanded polar sensing reduces these sampling biases. Monthly variations also fold in storm-rich months during the SC25 maximum, complicating the attribution of peaks solely to equinox effects.

Validation against operational outcomes is incomplete. Many operators lack systematic, time-stamped logging of HF outages, GNSS integrity degradations, and avionics single-event effects aligned to space weather indicators; radiation exposure records are often delayed or aggregated, constraining feedback to thresholds and training. As incident-linked evidence accumulates, MOD/SEV trigger levels could be recalibrated to reflect observed impact severity and frequency (e.g., HF un-usability duration/coverage, GNSS continuity/integrity excursions, measured dose rates). If impacts routinely occur below current MOD triggers, they could be lowered or refined; if SEV alerts yield negligible disruption, tighten definitions or raise triggers to reduce over-warning and align with agreed risk tolerances. Finally, public communication remains uneven: messaging can oscillate between hype and silence, while aviation needs concise, impact-based briefings that explain mitigations and uncertainty consistently across providers.

Finally, the advisory record is shaped by operational thresholds and evolving practice. First, issuance is gated by ICAO MOD/SEV thresholds, so the dataset systematically under-represents sub-threshold disturbances that may still produce local operational effects (especially for GNSS scintillation and HF usability). Second, advisory volumes reflect heterogeneous observing networks, model stacks, and centre workflows, which affect whether thresholds are crossed in a given grid cell and hour; inter-centre differences should therefore not be interpreted as service quality differences without controlling for inputs and algorithms. Third, the service matured during the study interval, and internal harmonisation guidance introduced in 2025 reduces stylistic and timing differences but cannot remove bias introduced by uneven sensor coverage and model choices. Fourth, the 30° × 15° geographic grid and coarse region codes compress spatial structure and can overstate affected areas when impacts are localised (e.g., equatorial post-sunset scintillation patches or auroral arcs). These limits constrain generalisation beyond the operational question the dataset answers: 'where and when were ICAO thresholds met and communicated', rather than 'where and when did the underlying geophysical disturbance occur'.

4. Conclusions and recommendations

Civil aviation is measurably exposed to space weather variability along four principal pathways: short-wave radio absorption from flare or particle-driven D-region ionisation, depletion of the F layer after geomagnetic storms resulting in MUF depression, small-scale ionospheric structuring that degrades GNSS accuracy, integrity, and availability, and radiation environments at flight altitudes during SEP events. During SC25, the rise from the 2019 minimum to the broad maximum in late-2024 to early-2025 increased the frequency of drivers, and advisory production followed this decadal envelope. Month-of-year aggregates show equinox-biased peaks that are consistent with enhanced solar-wind–magnetosphere coupling, while specific storm months in 2024–2025 amplify the counts beyond climatology. Geographic footprints concentrate in the low-latitude EIA with a pronounced maximum over South America for GNSS advisories, and in auroral and polar-cap belts for HF, with a Northern Hemisphere excess in this period. Centre-to-centre differences in issued volumes primarily reflect heterogeneous observing networks, modelling stacks, and workflow logic rather than inconsistencies with ICAO formats [7, 13–15, 20, 24, 25, 29].

Table 2 groups the recommendations by stakeholder and time horizon to support implementation planning and make responsibilities explicit.

Actionable priorities follow at three levels.

At the operational level, airlines and ANSPs should institutionalise trigger-to-action tables aligned with Annex 3 and Doc 10100 for HF COM, GNSS and RADIATION, with explicit ties to measurable indicators such as regional absorption in decibels at 10–15 MHz, ROTI and scintillation indices S4 and $\sigma\varphi$, augmentation status messages, and integral proton fluxes with modelled $H^*(10)$. These tables should be embedded in flight planning and flight-watch tooling, supported by recurrent training that rehearses minutes-to-days sequencing of impacts and includes procedures for rare but consequential solar radio bursts that affect surveillance. Operators should maintain systematic, time-stamped logs of HF outages, GNSS integrity degradations, and avionics anomalies to enable outcome-based verification and post-event learning [20, 24, 41, 42].

At the technical level, investment is needed in high-latitude sensing, where aviation exposure is greatest and present coverage is sparse. Priorities include denser riometer and GNSS networks, selected ionosondes in the Southern Ocean and North Pacific sectors, assimilation of riometer and HF measurements into absorption models used for advisory decisions, and fleetborne, low-latency radiation detectors to support rapid SEP onset assessment and $H^*(10)$ nowcasting. For navigation, real-time monitoring of GNSS augmentation performance and route-corridor ROTI, S4, and $\sigma\varphi$ should feed alerting logic. For avionics, qualification and maintenance programmes should include particle-effect testing appropriate to flight altitudes and standardised anomaly logging to support cross-fleet analysis of single-event effects [20, 27, 28, 43]. Given the coarse $30° \times 15°$ ICAO grid, centres should complement advisories with higher-resolution overlays where feasible and publish uncertainty ranges for affected extents.

At the governance level, sustained funding for the four Space Weather Centres and for underpinning ground and space observations is essential. The WMO ET-SWx framework should extend verification guidance that links advisory thresholds to operational outcomes such as raised minima, diversions and communication

Stakeholder	Short-term (0 to 12 months)	Medium-term (1 to 3 years)	Long-term (3+ years)
Airlines and operators	Implement trigger-to-action rules for HF COM, GNSS, and radiation, and link each action to the advisory identifier in dispatch logs [22].	Integrate space weather triggers into route and fuel planning, and perform post-event reviews using operational logs (delays, diversions, communication, and navigation performance) [3].	Embed space weather risk into network resilience and sustainability metrics (avoidable reroute rate, fuel penalty under space weather constraints) to guide long-term investment and procedures [22].
ANSPs and airports	Publish and rehearse procedural fallbacks for GNSS degradation and communication constraints, including capacity measures for reduced navigation performance [4].	Expand monitoring of GNSS integrity and availability at critical airports and corridors; incorporate space weather scenarios into joint exercises with operators. [4]	Build resilient concepts of operations that remain safe when GNSS is partially unavailable for multi-day periods, and ensure alternative navigation options remain available [4].
SWXCs and MET authorities	Standardise advisory interpretation and traceability (updates, replacements, cancellations) and provide user-facing guidance that maps advisory categories to operational mitigations [22].	Adopt harmonised verification metrics tied to advisory performance and user outcomes; improve machine-readable dissemination and stable archives [22].	Sustain aviation-impact services, including radiation dose-rate support and continuous improvement loops based on verified outcomes [4].
Regulators and governance (ICAO and WMO)	Clarify minimum expectations for trigger-to-action use, and encourage standardised reporting of operational outcomes to build the evidence base [22].	Define outcome-linked KPIs and support data-sharing agreements that enable independent evaluation while protecting sensitive operations data [22].	Institutionalise resilience planning for rare, high-impact events and periodic review of thresholds as services mature [4].

Table 2.
Operational recommendations grouped by stakeholder and time horizon (aligned with demonstrated continuity and efficiency impacts).

outages, and coordinate routine inter-comparison exercises across centres to benchmark models and workflows. The Guidance for the Issuance of Space Weather Advisories has reduced stylistic differences; the next step is common performance metrics, including timeliness, probability of detection and false-alarm rate conditioned on logged outcomes and season. Data-sharing agreements should enable anonymised exchange of HF outage reports, GNSS integrity events, avionics anomaly logs, and in-flight dose measurements to improve thresholds and models. Finally, risk communication for aviation should adopt concise, impact-based messaging with transparent uncertainty statements to avoid both under-reaction and over-reaction during active periods.

Taken together, these recommendations provide a practical route to reducing safety and efficiency risks throughout the active years of SC25. They align scientific understanding of solar-terrestrial drivers with decision-ready indicators, strengthen the evidence base through targeted observations and logging, and harmonise services so that similar conditions prompt similar actions across fleets and regions.

Author details

Judith de Patoul[1*], Tobias G.W. Verhulst[2] and Yana Maneva[1]

1 Solar Terrestrial Centre of Excellence - SIDC, Royal Observatory of Belgium, Brussels, Belgium

2 Royal Meteorological Institute of Belgium and Solar Terrestrial Centre of Excellence, Brussels, Belgium

*Address all correspondence to: Judith.dePatoul@oma.be

References

[1] COST Action. Developing the Scientific Basis for Monitoring, Modelling and Predicting Space Weather. Vol. **724**, Brussels: European Cooperation in Science and Technology; 2008

[2] Lilensten J, Belehaki A, Messerotti M, Vainio R, Waterman J, Poedts S. Developing the scientific basis for monitoring, modelling and predicting space weather, OPOCE, COST 724 final report. 2009

[3] Abt Associates. Economic Impact of Space Weather on the United States. Cambridge, MA: Abt Associates; 2017

[4] National Research Council. Severe Space Weather Events: Understanding Societal and Economic Impacts. Washington, DC: National Academies Press; 2008

[5] Royal Academy of Engineering. Extreme Space Weather: Impacts on Engineered Systems and Infrastructure. London: Royal Academy of Engineering; 2013

[6] Cabinet Office. National Risk Register of Civil Emergencies. London: UK Cabinet Office; 2017

[7] Temmer M. Space weather: The solar perspective. Living Reviews in Solar Physics. 2021;**18**:4. DOI: 10.1007/s41116-021-00030-3

[8] Patel BD, Joshi B, Cho KS, Kim RS. DH type II radio bursts during solar cycles 23 and 24: frequency-dependent classification and their flare-CME associations. Solar Physics. 2021;**296**:142. DOI: 10.1007/s11207-021-01890-6

[9] Grandin M, Aikio AT, Kozlovsky A. Properties and geoeffectiveness of solar wind high-speed streams and stream interaction regions during solar cycles 23 and 24. Journal of Geophysical Research: Space Physics. 2019;**124**:3871–3892. DOI: 10.1029/2018JA026396

[10] Temmer M, Veronig AM, Vršnak B, Rybák J, Gömöry P, Stoiser S, Maričić D. Acceleration in fast halo CMEs and synchronized flare HXR bursts. The Astrophysical Journal Letters. 2008;**673**: L95–L98. DOI: 10.1086/527414

[11] Gonzalez WD, Tsurutani BT, Clúa de Gonzalez AL. Interplanetary origin of geomagnetic storms. Space Science Reviews. 1999;**88**:529–562. DOI: 10.1023/A:1005160129098

[12] Desai M, Giacalone J. Large gradual solar energetic particle events. Living Reviews in Solar Physics. 2016;**13**:3. DOI: 10.1007/s41116-016-0002-5

[13] Fiori RAD, Trichtchenko L, Balch C, Spanswick E, Groleau S. Characterizing auroral-zone absorption based on global Kp and regional geomagnetic hourly range indices. Space Weather. 2020;**18**: e2020SW002572. DOI: 10.1029/2020SW002572

[14] Fiori RAD, Koustov AV, Chakraborty S, Ruohoniemi JM, Danskin DW, Boteler DH, Shepherd SG. Examining the potential of the Super Dual Auroral Radar Network for monitoring the space weather impact of solar X-ray flares. Space Weather.

2018;**16**:1348-1362. DOI: 10.1029/2018SW001905

[15] Fiori RAD, Danskin DW, Pitteway MLV, Sauter E, Wilton A, Jayachandran PT. Occurrence rate and duration of space-weather impacts on high-frequency radio communication used by aviation. Journal of Space Weather and Space Climate. 2022;**12**:21. DOI: 10.1051/swsc/2022017

[16] Warnant R, Foelsche U, Aquino M, Bidaine B, Gherm V, Hoque MM, Kutiev I, Lejeune S, Luntama JP, Spits J, Strangeways HJ, Wautelet G, Zernov N, Jakowski N. Mitigation of ionospheric effects on GNSS. Annals of Geophysics. 2009;**52**(3-4):373–390. DOI: 10.4401/ag-4585

[17] Jakowski N, Mayer C, Hoque MM, Wilken V. Total electron content models and their use in ionosphere monitoring. Radio Science. 2011;**46**: RS0D18. DOI: 10.1029/2010RS004620

[18] Dyer CS, Lei F, Clucas SN, Smart DF, Shea MA. Calculations and observations of solar particle enhancements to the radiation environment at aircraft altitudes. Advances in Space Research. 2003;**32** (1):81–93. DOI: 10.1016/S0273-1177(03)90374-7

[19] Berdermann J, Kriegel M, Banyś D, Heymann F, Hoque MM, Wilken V, Borries C, Heßelbarth A, Jakowski N. Ionospheric response to the X9.3 flare on 6 September 2017 and its implication for navigation services over Europe. Space Weather. 2018;**16**:1604-1615. DOI: 10.1029/2018SW001933

[20] Kauristie K, Andries J, Beck P, Berdermann J, Berghmans D, et al. Space weather services for civil aviation: Challenges and solutions. Remote Sensing. 2021;**13**(18):3685. DOI: 10.3390/rs13183685

[21] Marqué C, Klein KL Monstein C, Opgenoorth H, Pulkkinen A, Buchert S, Krucker S, Van Hoof R, Thulesen P. Solar radio emission as a disturbance of aeronautical radionavigation. Journal of Space Weather and Space Climate. 2018;**8**:A42. DOI: 10.1051/swsc/2018029

[22] Eastwood JP, et al. The economic impact of space weather: Where do we stand? Risk Analysis. 2017;**37**(2):206–218. DOI: 10.1111/risa.12765

[23] International Civil Aviation Organization. Annex 3 to the Convention on International Civil Aviation: Meteorological Service for International Air Navigation. 20th edn. Montreal: ICAO; 2018. https://elibrary.icao.int/product/264207

[24] International Civil Aviation Organization. Manual on Space Weather Information in Support of International Air Navigation. 1st edn. Montreal: ICAO; 2019. Doc 10100. https://store.icao.int/en/manual-on-space-weather-information-in-support-of-international-air-navigation-doc-10100

[25] Hathaway DH. The solar cycle. Living Reviews in Solar Physics. 2010;**7**:1. DOI: 10.12942/lrsp-2010-1

[26] Giersch O, et al. Solar radio burst statistics and implications for space weather effects. Space Weather. 2017;**15**:1511–1522

[27] Sadykov VM, Kitiashvili IN, Tobiska WK, Guhathakurta M. Radiation Data Portal: integration of radiation measurements at aviation altitudes and solar-terrestrial environment observations. Space

Weather. 2021;**19**:e2020SW002653. DOI: 10.1029/2020SW002653

[28] Tobiska WK, Bouwer D, Smart D, Shea M, Bailey J, Didkovsky L, Judge K, Garrett H, Atwell W, Gersey B, et al. Global real-time dose measurements using the Automated Radiation Measurements for Aerospace Safety (ARMAS) system. Space Weather 2016;**14**:1053–1080. DOI: 10.1002/2016SW001419

[29] Aa E, Zou S, Eastes R, Karan DK, Zhang SR, Erickson PJ, Coster AJ. Coordinated ground-based and space-based observations of equatorial plasma bubbles. Journal of Geophysical Research: Space Physics. 2020;**125**: e2019JA027569. DOI: 10.1029/2019JA027569

[30] Millas D, Maneva Y, Shukhobodskaia D, de Patoul J. Impact of recent geomagnetic storms on aviation as observed by PECASUS. Talk presented at UKSWSE III, Sheffield, Space Pole Publications Server, CTALK-2025-0099. 2025. Available at: https://publi2-as.oma.be/record/7646?ln=en [Accessed: 2025-September-11]

[31] International Civil Aviation Organization. NAT Space Weather Exercise Dec 2025. Presentation to Joint SAT IMG/7 and SAT SOG/7, SAT-IMG7-PPT-02J. ICAO. 2026. Available at: https://www.icao.int/sites/default/files/WACAF/MeetingDocs/2026/Joint%20SAT%20IMG7%20%26%20SAT%20SOG%207/WORKING%20PAPERS/SAT-IMG7-PPT-02J-NAT-SWX2025-summary.pdf

[32] Lockwood M, Owens MJ, Barnard LA, Haines C, Scott CJ, McWilliams KA, Coxon JC. Semi-annual, annual and Universal Time variations in the magnetosphere and in geomagnetic activity: 1. Geomagnetic data. Journal of Space Weather and Space Climate. 2020;**10**:23. DOI: 10.1051/swsc/2020023

[33] Russell CT, McPherron RL. Semiannual variation of geomagnetic activity. Journal of Geophysical Research. 1973;**78**(1):92–108. DOI: 10.1029/JA078i001p00092

[34] Appleton E. Two Anomalies in the Ionosphere. Nature. 1946;**157**:691. DOI: 10.1038/157691a0

[35] Balan N, et al. A brief review of the equatorial ionization anomaly and ionospheric irregularities. Frontiers in Astronomy and Space Sciences. 2018;**5**:32

[36] Lu G, Huba JD, Valladares C. Modeling ionospheric super-fountain effect based on the coupled TIMEGCM-SAMI3. Journal of Geophysical Research: Space Physics. 2013;**118**:2527–2535. DOI: 10.1002/jgra.50256

[37] Tsurutani B, et al. Global dayside ionospheric uplift and enhancement associated with interplanetary electric fields. Journal of Geophysical Research. 2004;**109**:A08302

[38] Rishbeth H, Mendillo M. Patterns of F2-layer variability. Journal of Atmospheric and Solar-Terrestrial Physics. 2001;**63**(15):1661–1680. DOI: 10.1016/S1364-6826(01)00036-0

[39] Astafyeva E, Bagiya MS, Förster M, Nishitani N. Unprecedented hemispheric asymmetries during a surprise ionospheric storm: a game of drivers. Journal of Geophysical Research: Space Physics. 2020;**125**: e2019JA027261. DOI: 10.1029/2019JA027261

[40] Basu S, MacKenzie E, Basu S. Ionospheric constraints on VHF/UHF communications links during solar maximum and minimum periods. Radio Science. 1988;**23**(3):363–378. DOI: 10.1029/rs023i003p00363

[41] Bain HM, Copeland K, Onsager TG, Steenburgh RA. NOAA Space Weather Prediction Center radiation advisories for the International Civil Aviation Organization. Space Weather. 2023;**21**:e2022SW003346. DOI: 10.1029/2022SW003346

[42] Civil Aviation Authority. Impacts of Space Weather on Aviation. CAP1428, Version 2. Gatwick: Civil Aviation Authority. 2020. Available at: https://www.caa.co.uk/data-and-publications/publications/documents/content/cap1428/

[43] Roy A, Sarkar R, Lee C. Extensive study of radiation dose on human body at aviation altitude through Monte Carlo simulation. Life Sciences in Space Research. 2021;**31**:1–13. DOI: 10.1016/j.lssr.2021.07.001

[44] Hargreaves JK. The Solar-Terrestrial Environment. Cambridge, UK: Cambridge University Press; 1992

[45] Prölss GW. Ionospheric F-region storms. In: Volland H, editor. Handbook of atmospheric electrodynamics. 2nd ed. Boca Raton, Fla: CRC Press; 1995. p. 195–235

[46] Kavanagh AJ, Kosch MJ, Honary F, Senior A, Marple SR, Woodfield EE, McCrea IW. The statistical dependence of auroral absorption on geomagnetic and solar wind parameters. Annales de Geophysique. 2004;**22**(3):877–887

[47] Xue D, Wu L, Xu T, Wu CL, Wang Z, He Z. Space weather effects on transportation systems: a review of current understanding and future outlook. Space Weather. 2024;**22**: e2024SW004055. DOI: 10.1029/2024SW004055

[48] Maneva Y. ICAO Space Weather Information Service (SWIS): Education and Guidance Materials. In: Presentation at METP SWX User Workshopon behalf of the Space Weather Coordination Group. International Civil Aviation Organization; 2025. Rome: 20 October 2025. https://www.icao.int/sites/default/files/METP/Event%20and%20Workshop/SWX%20User%20Workshop%202025/METP-SWX-User-Workshop-SP02-SWIS-Education-and-Guidance-Material-new.pdf

[49] Rishbeth H, Müller-Wodarg ICF. Why is there more ionosphere in January than in July? The annual asymmetry in the F2-layer. Annales Geophysicae. 2006;**24**:3293–3311. DOI: 10.5194/angeo-24-3293-2006

Chapter 4

Optimising Fuel Cell Systems for Aviation

Caroline Willich, Tobias Graf, Daniel Frank, Roland Ademi, Samara Brandão de Almeida and Christiane Bauer

Abstract

Commercial fuel cell systems are already used in ground applications, but in aviation, they require adaptations to hardware and software, as the system operates under varying ambient pressure during flight and faces strict weight and space constraints. Fuel cell systems optimised for aviation have yet to be developed. The chapter outlines challenges and examines necessary modifications to an FC system designed for ground applications to optimize it for providing propulsive power in an aircraft. It also gives an outlook for further optimisation potential for the use of fuel cells in aviation. Low ambient pressures decrease the performance of fuel cells but also reduce outlet pressure and maximal mass flow of a compressor, which can be used for mitigating low pressures. It is shown how the compressor requires energy, which reduces the system power, and how it may not provide sufficient mass flow if not optimised for the target cruise height, limiting the output of the fuel cell at higher currents. An approach for developing a control strategy ensuring the safe and stable dynamic operation despite varying inlet conditions is presented, and further optimisation potentials, including pressure-dependent temperature and water management for optimal membrane humidification, the integration of a turbine for recuperation, the use of High-Temperature-PEM, and the integration with liquid hydrogen storage, are addressed.

Keywords: PEM fuel cell, fuel cell system optimisation, varying operating pressure, sustainable aviation, hydrogen propulsion

1. Introduction

Using hydrogen as a fuel for propulsion in aviation is considered a promising option to decrease the climate impact of aviation [1]. Hydrogen has a much higher specific energy of 33 kWh/kg in comparison to batteries (0.06 – 0.27 kWh/kg for state-of-the-art lithium-ion batteries [2]). Using hydrogen as energy storage therefore allows for longer flight distances. Operating aircraft with hydrogen turbines is predicted to decrease the climate impact by 75% [1]. Converting hydrogen in fuel cells and using the electrical energy that is produced in an all-electric aircraft instead can even decrease climate-relevant emissions by up to 90% [1]. Fuel cells convert chemically bound energy directly into electricity. They are therefore not limited by the Carnot efficiency and offer high efficiencies, and when operating fuel cells with pure hydrogen, the only reaction product is water.

Several hydrogen-powered planes have already flown. Boeing demonstrated a 20-minute horizontal flight powered by a hydrogen fuel cell in 2008. A battery provided additional power for take-off and climb [3]. The Antares DLR-H2, developed by the German Aerospace centre flew powered only by a fuel cell in 2009, and in 2016, 2020, and 2022, the HY4 flew repeatedly [4, 5]. The company ZeroAvia reported a hydrogen-powered flight with a Piper M-class in 2020 [6].

Despite the advantage of a high specific energy of hydrogen, the specific power of fuel cell systems is a major challenge. In comparison to conventional turbines used in propulsion systems, which have a specific power of around 25 kW/kg [7], fuel cell technology is still very heavy, with the specific power of commercial automotive systems being in the range of 0.7 kW/kg [8] to 0.86 kW/kg [9] at the system level. Optimising fuel cell system performance while minimizing system weight is therefore essential for the successful implementation of the technology in aviation. It is expected that power densities of up to 3.5 kW/kg will be achievable by 2050 [10] using low-temperature PEM technology. Besides the weight challenge, the operation of fuel cell systems at high altitudes offers additional challenges in comparison to applications on the ground.

2. Challenges

PEM fuel cell systems are already commercially available for applications in the mobility sector on the ground. The use in aircraft, however, is not yet state of the art, and several additional challenges have to be met. For aviation, requirements like weight, space, high efficiency, reliability, and safety are even more rigorous than in other mobile applications. System weight directly impacts the energy required for propulsion during a flight, and as the maximum take-off weight of an aircraft is limited, a surplus in system weight will lead to a lower payload mass, resulting in fewer available passenger seats and higher cost-per-passenger-kilometre. Additionally, safety regulations in aviation are stricter than in the automotive sector, with all components and subsystems having to meet certification standards for aviation.

2.1 Operation at altitude

The operation at high altitudes, at which airplanes travel, directly leads to additional challenges. Ambient pressure and temperature decrease significantly at high altitudes. Up to an altitude of 11 km, the International Standard Atmosphere [11] indicates a temperature decline of - 6.5 K per km, while the temperature in the tropopause (at an altitude between 11 and 20 km) remains constant.

The cruise altitude for commercial aircraft is between 10 km (33,000 feet) and 13 km (42,000 feet), while recreational aircraft fly lower at 13,000–15,000 feet (4–4.5 km) [12]. Any components placed in non-insulated and non-pressurised parts of the airplane are exposed to those conditions. During start, the ambient temperatures can reach the highest temperatures on Earth (58°C measured in El Azizia, Libya, in 1922 [13]), while during cruise, very low temperatures are to be expected (around - 56.5°C at an altitude of 11,000 m). The ambient pressure during cruise will be around 0.2 bar at this altitude.

Especially the low pressures pose a challenge for operating fuel cells with ambient air at high altitudes because low operational pressures at the fuel cell stack decrease its performance [14]. Operational pressures influence not only the open circuit voltage but also the activation overpotential. At lower partial pressures, fewer reactants are adsorbed on the active sites within the electrodes, decreasing the charge transfer reaction as well as the exchange current density [15]. This loss depends strongly on pressure and is considered to have the strongest influence on pressure-related voltage changes [16]. Concentration losses are also influenced by pressure since the effective diffusion coefficient is pressure-dependent [17] and the limiting current density decreases [18]. The air flow velocity along the gas channels of the cell is also affected by pressure, which can influence the local availability of reactants at the cell and increase the risk of dehydration of the membrane. Pressure also has an influence on the water management of the cell, as low pressure also affects the humidity of the cathode air. The dew point temperature decreases with lower pressures [19], and for the same amount of water in the air, the relative humidity decreases. For the membranes that are commonly used in PEM fuel cells, lower relative humidity reduces the membrane's H^+ conductivity and therefore increases its ohmic resistance [20], while too much water leads to flooding of the cell. Werner et al. [14] found that the overall stack performance and efficiency decrease by 5.3% when the operating pressure was lowered from 950 to 700 mbar.

3. System aspects

A fuel cell stack has to be supplied with fuel (H_2) on the anode and with air or oxygen on the cathode side, as is shown schematically in **Figure 1**. The supply of H_2 is controlled with the help of valves. Recirculation increases fuel usage, so less H_2

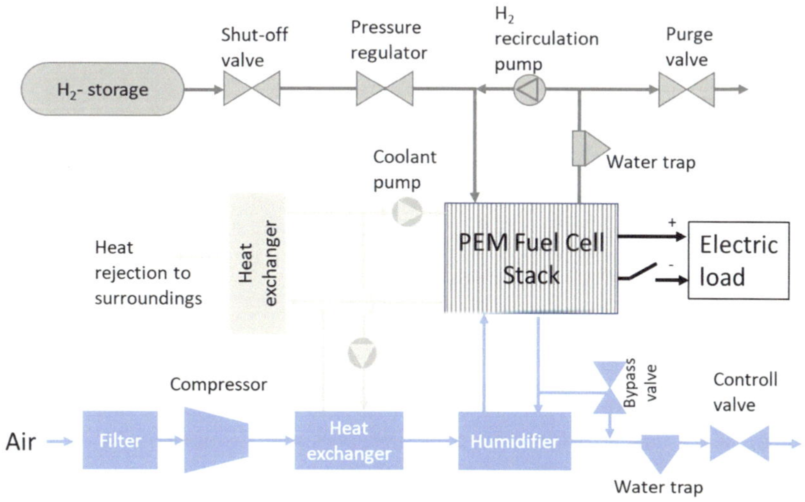

Figure 1.
Simplified schematic architecture of a PEM fuel cell system. Adapted from [21].

leaves the system through the purge valve. The air is taken from the surroundings. A filter, compressor, intercooler, and humidifier ensure that it enters the fuel cell stack at the desired conditions. In addition to the media supply, the fuel cell needs a suitable cooling system to control the stack temperature and an electrical connection to the load.

Under some circumstances, fuel cell systems can be operated without humidifying the gases, but usually, the reactant gases are humidified to reduce degradation of the cells [22–24]. Humidification is done passively via membrane humidifiers, as shown in **Figure 1,** or actively via water injection. In the case of liquid water injection, the evaporation also cools the gas stream. If cooling is required at this point in the system, this can potentially eliminate the need for an intercooler. It is also possible to replace the control valve at the exit of the exhaust air with a turbine to recuperate energy, as will be addressed in Chapter 4.1.

3.1 Matching of compressor and fuel cell

To mitigate the influence of low pressure, the air for the cathode of the fuel cell system can be pressurised. In an all-electric airplane, it stands to reason to employ an electrical compressor, which can be electrically supplied by the fuel cell system itself. Different compressor technologies can be used [25], but most developments focus on centrifugal compressors [26–28] as they are comparably light weight and suited for the mass flows and pressure ratios required for typical fuel cell system sizes.

Pressurising fuel cell systems is already done for applications on the ground [29, 30]. In aviation, higher pressure ratios will be required, depending on the desired target operating pressure, due to the low inlet pressures at high altitudes. To be able to operate the fuel cell at 1 bar when ambient air is at 0.25 bar will already require a compressor ratio of more than 4, as pressure losses in the air path components (filter, thermal, and humidity management) need to be compensated. Commercial systems for ground applications usually operate below pressures of 3–4 bar [31], as a further increase in pressure does not result in a significant increase in power. Many only operate with slightly increased pressures [21].

Figure 2 shows the increase in stack power for operation at different pressures above 1 bar for a stack, as well as the required compressor power for providing the higher pressure on the ground and at altitudes of 5,000 m, 10,000 m, and 15,000 m. The pressure-dependent fuel cell stack data is based on the manufacturer [32] and literature data [33] of an NM5 fuel cell stack produced by ElringKlinger containing 300 cells.

To obtain the system power, the compression power P_{el} is calculated according to Eq. (1) [25] and subtracted from the stack power.

$$P_{el} = P_{is} \cdot \frac{1}{\eta} = c_p \cdot T_{in} \cdot \dot{m} \cdot \left[\left(\frac{p_{out}}{p_{in}} \right)^{\frac{\kappa-1}{\kappa}} - 1 \right] \cdot \frac{1}{\eta} \quad (1)$$

The efficiency η in Eq. (1) was assumed to be 0.6, the specific heat capacity c_p at constant pressure was considered a constant, and, κ is the isentropic exponent was used. The inlet temperature T_{in} and inlet pressure p_{in} are assumed to be the temperature and pressure at the corresponding altitude, as given by the International Standard Atmosphere, as shown in **Table 1**. The mass flow $\dot{m}$ is obtained from the

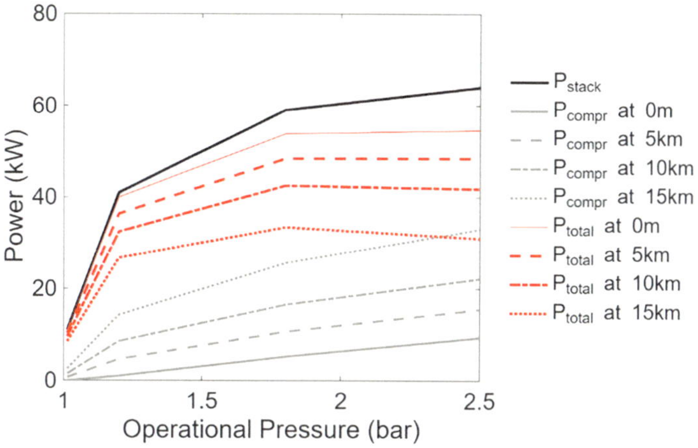

Figure 2.
Power output of a fuel cell stack, required compression power, and overall system output. Fuel cell data based on [32–34].

Altitude [km]	Pressure [bar]	Temperature [K]
0	1.0132	288
5	0.5402	256
10	0.2644	223
15	0.1204	216

Table 1.
Altitude, ambient pressure, and temperature as defined by the International Standard Atmosphere [11].

current drawn and the cathode stoichiometries given by the cited sources as 1.7 for the 1 and 2.5 bar curves, and between 1.7 and 2.7 for the 1.8 bar curve [34].

It can be seen from **Figure 2** that, above an operating pressure of the fuel cell of 1.8 bar, the overall system power does not increase. It follows that an operational pressure above approximately 1.8 bar offers no benefit for the considered fuel cell data. Especially at high altitudes, the increasing power demand of the compressor even leads to a slightly decreasing system output above 1.8 bar operational pressure.

In a real application, the operational point of the fuel cell will be defined depending on the power demand, which is highest during take-off. During cruise, less power (in the order of 50%) is required. It can, therefore, offer a benefit to operate the system with a varying pressure ratio and design the fuel cell system to operate at higher pressures during take-off, where inlet pressures are higher anyway, and settle for a lower operational pressure during cruise, where inlet pressures are lower but less power is needed. A fuel cell operated with a compression ratio of 1.8 will operate at a pressure of 1.8 bar (minus pressure losses) and yield a system power of 54 kW on the ground, according to **Figure 2**. Assuming roughly half of that power is required for cruise, a compression ratio of almost 4.5 would be required to reach an operational pressure of close to 1.2 bar to obtain approximately 32 kW at an altitude of 10 km, as can be seen from **Figure 2**.

3.1.1 Limits of the compressor and influence of ambient conditions

The operational map of a compressor, as seen in **Figure 3,** is limited by the surge line on the left, towards the top-right by the maximum rotational speed or maximum available torque, towards the bottom-right by the choke limit and throttle line of the downstream system, and towards the bottom by minimal compressor speed and – depending on the bearing type – minimal bearing pressure [35].

The maximum outlet pressure and mass flow that a compressor can deliver depend on its inlet conditions, as can be seen in **Figure 3**. For a given compressor, the outlet pressures will be lower at lower inlet pressures, and the maximum mass flow also decreases. This effect needs to be considered when operating a fuel cell system at high altitude, and a suitable compressor needs to be selected.

That the compressor of a fuel cell system developed for operation on the ground should be changed when adapting the system to aviation has been shown by Willich et al. [21]. A commercial, slightly modified system was analysed in a low-pressure environment. Details of the experimental setup and measurement uncertainty are given in [25]. When operating the system at lower pressures, the compressor increases its speed to supply the system with the required mass flow despite the lower air density. However, once the compressor speed reaches its limit of 120 krpm, the air mass flow decreases at lower pressures, leading to a lower cathode stoichiometry and superimposing a starvation effect over the purely pressure-related performance decrease.

Figure 4 shows stack as well as system power, compressor speed, cathode inlet pressure, and air mass flow plotted against current for measurements at ambient pressures of 940 mbar, 900 mbar, 845 mbar, 795 mbar, and 750 mbar. The inlet temperature was 15°C for all measurements. It is observable that the power at constant current decreases with altitude, as expected. The maximum possible power also decreased with pressure. It can also be observed that the required compressor power increases with current. The compressor power decreases the overall system power. The fuel cell outlet temperatures varied less than 3 K between the

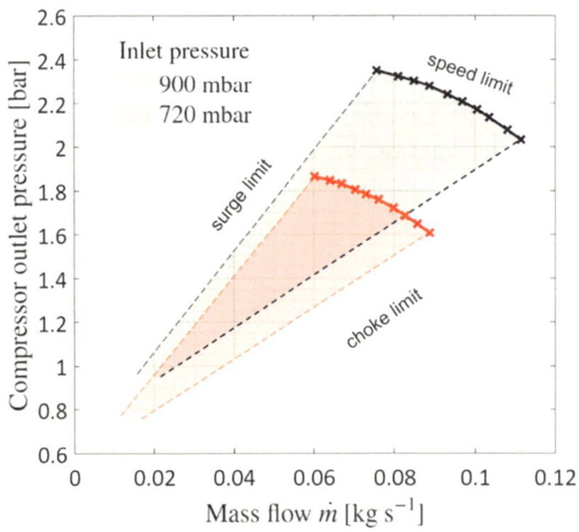

Figure 3.
Operational maps of a Fischer EMTC 120 K compressor (Fischer Fuel Cell Compressor AG) measured at 900 mbar and 720 mbar ambient pressure and 15°C ambient temperature (own measurement).

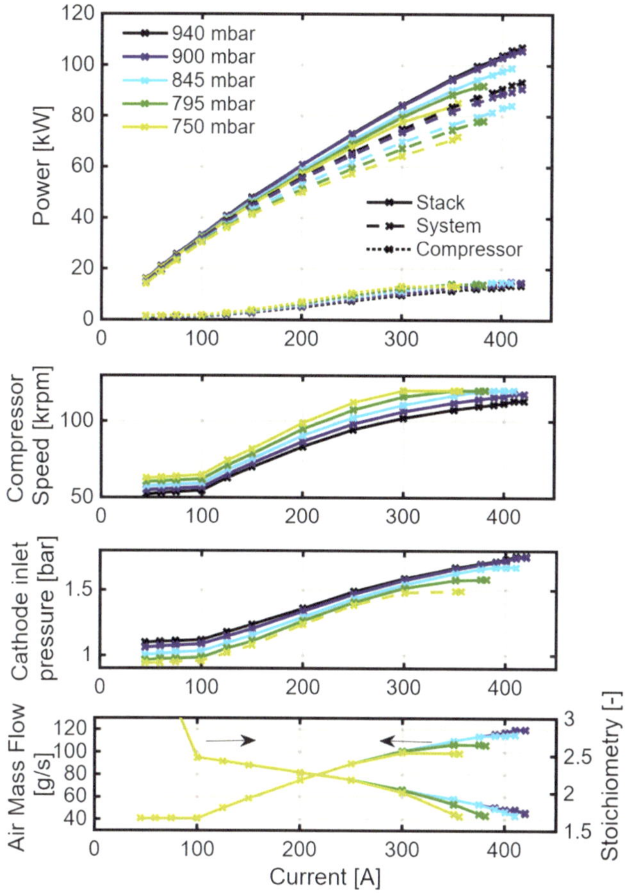

Figure 4.
Stack and system power, compressor speed, cathode inlet pressure, air mass flow, and stoichiometry against current at different altitudes that correspond to ambient pressures of 940 mbar, 900 mbar, 845 mbar, 795 mbar, and 750 mbar. Based on [21].

measurements; therefore, a temperature influence can be excluded. The power at high currents for 940 mbar and 900 mbar is similar because the compressor manages to compensate for the lower ambient pressure by increasing its speed and provides almost the same pressure and the same mass flow. However, at lower ambient pressures and high currents, the compressor can no longer compensate. As a consequence, lower inlet pressures lead to a decrease in air mass flow. It is clearly visible that at an inlet pressure of 750 mbar, the compressor reaches its speed limit at a fuel cell current of approximately 300 A, and for higher currents, the air mass flow can no longer be increased. This means that, in addition to lower pressures, starvation is superimposed on the characteristic curves of the examined fuel cell system.

The effect that a reduction in mass flow, due to the operating point of the compressor, can have is also observable in **Figure 5**. **Figure 5a**) shows simulated current-voltage curves for a fuel cell system based on the same system as shown in **Figure 4**. Within the system, a one-stage compressor (based on an EMTC 120k,

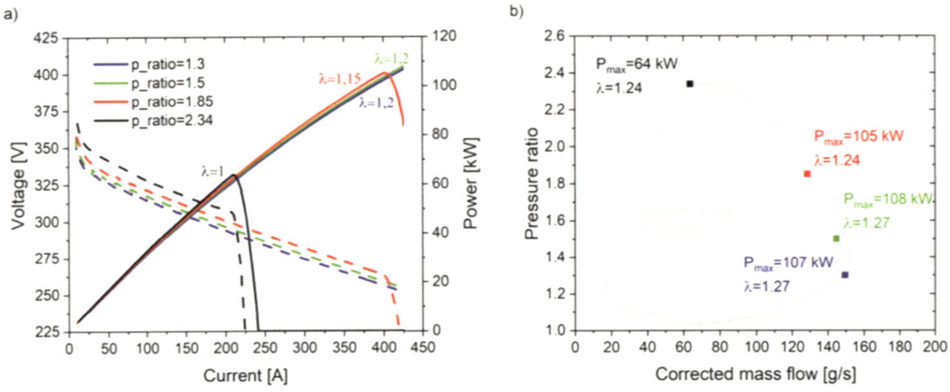

Figure 5.
Simulated characteristic curves for a fuel cell stack with a compressor operated at different setpoints. (a) Fuel cell characteristic curves (b) compressor map and setpoints [36].

Fischer Fuel Cell Compressor AG) is operated at different pressure ratio setpoints inside its operational map. The map and the operating setpoints are shown in **Figure 5b**), as well as the corresponding stoichiometry λ and maximum stack output power. Simulations were performed in MATLAB Simulink for ambient pressures of 0.54 bar. The maximum compressor speed was 110 krpm.

It can be seen that operation at higher pressures does not necessarily lead to a higher power output. The characteristic fuel cell curve at 2.34 bar only yields a higher output until a current of 225 A. Higher currents cannot be reached, as the compressor provides insufficient mass flow. The figure highlights the constraints imposed by the compressor map, where the achievable mass flow is limited by the pressure ratio.

It follows that, for the control of the compressor and its operating strategy, mass flow should be prioritised over pressure to prevent oxygen starvation, possible damage to the fuel cells, and a sudden decrease in power.

3.2 Control and stability

Besides the matching of the compressor and fuel cell, the dynamic operation of both has to be considered, and reliability has to be ensured, especially when fuel cell systems are used as the primary energy source for power trains in aviation. The fuel cell system has to be controlled to operate safely, and emergency shutdowns have to be avoided. For the control of the system, two important aspects have to be considered:

1. The cathode air mass flow demand must be met in all situations and should be prioritised over pressure increase to prevent low cathode stoichiometries and the consequent reduction in current, as shown above.

2. Unstable operating conditions must be avoided.

The compressor surge line marks a boundary between stable and unstable operation areas. In the compressor map, the unstable area is to the left of the surge line, where the backpressure is too high for the air mass flow. Crossing the surge line results in an oscillating flow, which can damage both the compressor and the fuel cell. Frank et al. [37] observed the compressor entering surge when transitioning from a setpoint with high air mass flow and high pressure to low air mass flow and low pressure. Frank et al. [37] consequently developed and implemented a surge control strategy without additional surge control valves, evacuating the air in the system to decrease through the pressure control valve before decreasing pressure, as can be seen in **Figure 6**.

Pressure and mass flow through the fuel cell can be controlled through the actuation of the compressor and a pressure control valve in the cathode exhaust. Both actuators influence mass flow and pressure, and changing either will always have an effect on both. When developing a control system, this has to be considered, and both actuators have to be coordinated to handle the coupling of mass flow and pressure. It has been shown that good control performance is possible for transient behaviour using an integral multi-variable approach by determining suitable control parameters and using the following considerations [38]:

To describe the operation of a centrifugal compressor independently of inlet pressure and temperature, the corrected mass flow $\dot{m}_{corr}$, the pressure ratio ψ and the corrected compressor speed N_{corr} is are commonly used [39]. The similarity values $\dot{m}_{corr}$ and ψ are inversely proportional to the compressor inlet pressure, as given in Eq. (2) and (3).

$$\dot{m}_{corr} = \dot{m}\frac{p_{ref}\sqrt{T_{in}}}{p_{in}\sqrt{T_{ref}}} \propto \frac{\dot{m}}{p_{in}} \tag{2}$$

$$\psi = \frac{p_{out}}{p_{in}} \tag{3}$$

If the inlet pressure decreases, the compressor speed has to be increased to retain the required mass flow and outlet pressure. The control response for the compressor speed is the same if the change in these similarity values is the same. This can be used for the development of a controller structure when the control errors are converted to corrected mass flows and pressure ratios as well. In this way, the parameter set of the control system can be independent of the ambient pressure [38].

Pressure, as well as mass flow, is influenced by the opening angle of the control valve, which is located between the fuel cell stack and the system's exhaust, as depicted in **Figure 1**. A smaller opening area throttles the mass flow, and therefore the pressure increases. The blocked area in the valve scales with sin (α), where α is the opening angle. The control signal can be linearised and the valve setpoint can be calculated as $\sin(\alpha)$. α can then be calculated afterwards [38].

Using the described relations, Frank et al. [38] observed good control performance and stable operation of the air supply system with this control system.

4. Further optimisation potential of fuel cells in aviation

A big challenge is the reduction of the weight of the fuel cell system and the required balance of plant by using lightweight materials and increasing the power-to -weight ratio of all components. A suitable packaging to decrease the required space is also necessary, for optimal integration into an aircraft. There are also possibilities to integrate the fuel cell system with other aviation functions. Heat and product

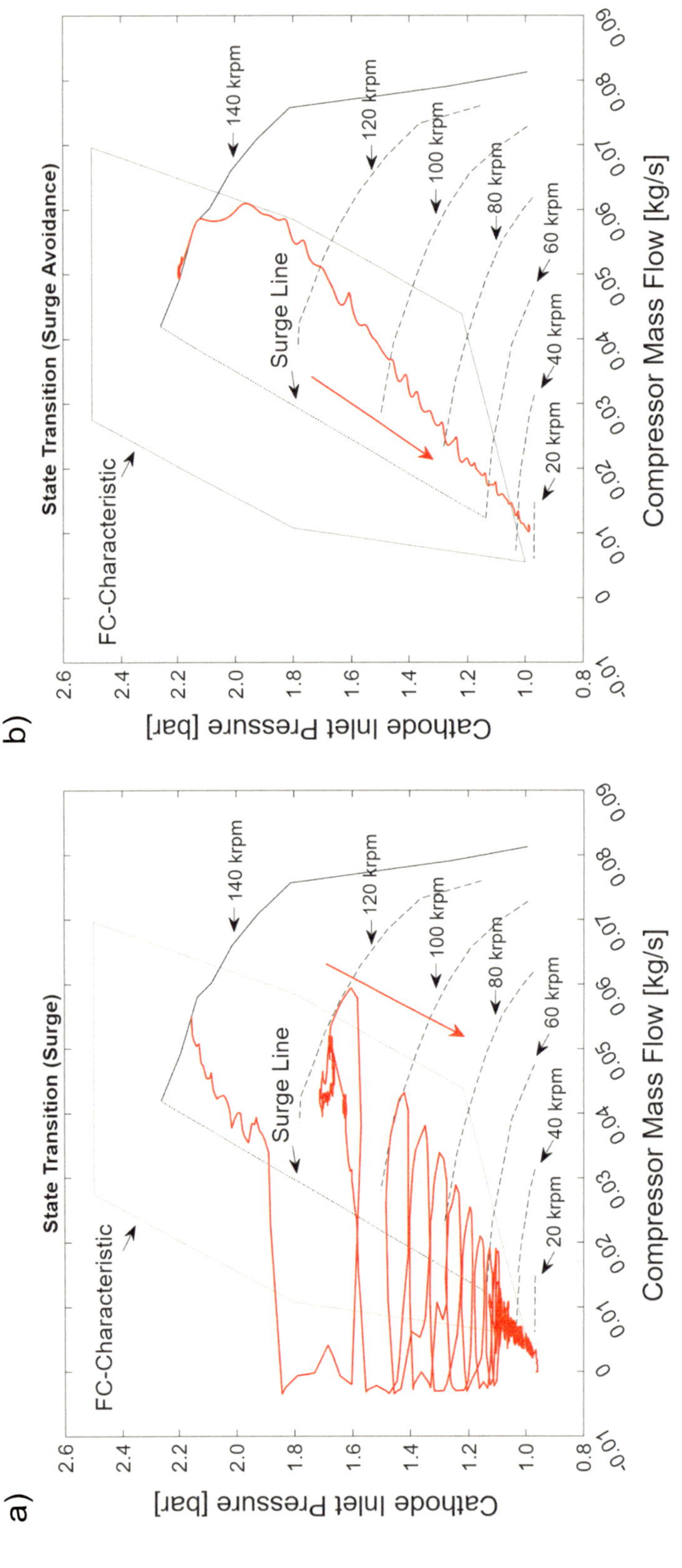

Figure 6.
Occurrence of compressor surge when changing from a setpoint at high pressure to one at low pressure (a), and the same transition with surge control implemented (b). Based on [37].

water, as well as the use of oxygen-depleted air, have been proposed for de-icing, water supply, or fire suppression [40–42]. Many of these will, however, increase complexity to an extent that makes them unfeasible or not viable. In the following chapters, several promising optimisation approaches are addressed.

4.1 Recuperation in turbine

A potential modification of the fuel cell system that can increase the overall power output is the use of a turbine in the cathode exhaust to recover energy from the cathode air, in addition to or instead of the control valve shown in **Figure 1**. The mass flow through the turbine is higher than in the compressor since the product water of the fuel cell reaction is added to the flow [27]. The amount of water varies according to the operating point of the fuel cell stack, which leads to a high variance in turbine mass flow and compressor power [27]. The inclusion of a turbine in the cathode exhaust has the potential to lower the specific energy consumption by up to approximately 24% [43] because between 36% and 43% of the compressor power can be supplied by the turbine [27]. Because of the high humidity in the cathode exhaust (close to 100% rH [44]) and the relatively low temperatures (<100°C), condensation is to be expected within the turbine, which influences turbine operation. Higher water-to-air ratios were found by Lück et al. [27] to shift the operating line of the turbine closer to the surge line, which needs to be considered in the control system to avoid unstable conditions. However, condensation did not significantly affect efficiency.

4.2 Humidity and temperature optimisation at varying pressures

As stated in Chapter 2.1, the operation of PEM fuel cells at lower pressures indirectly influences the ohmic resistance in the fuel cell due to the membrane humidity. Water management, therefore, is an important topic if the fuel cell's operating pressure is not increased by a compressor or is operated at varying pressures.

The partial pressure of water in air p_{H_2O} changes with temperature and pressure (Eq. [4]), but the saturation pressure p_{sat} (Eq. [5]) of water in air is almost independent of pressure [45]. It follows that, for a constant relative humidity rH as shown in Eq. (6), at low pressure p the water content x (Eq. (7)) is higher unless the temperature is reduced as well.

$$p_{H_2O} = f(p, T, x) \tag{4}$$

$$p_{sat} = f(T) \tag{5}$$

$$rH = \frac{p_{H2O}(p, T, x)}{p_{sat}(T)} = rH(p, T, x) \tag{6}$$

$$x = \frac{m_{H_2O}}{m_{dry\ air}} \tag{7}$$

In **Figure 7**, the mole fractions of water, N_2, and O_2 can be seen plotted against total pressure for a relative humidity of 80% at a temperature of 70°C on the left and for 50°C on the right. It can be seen that for a constant temperature, and rH the

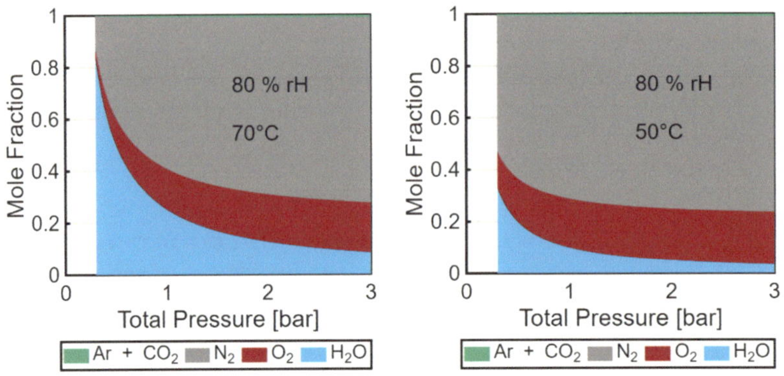

Figure 7.
Wet air composition at different pressure levels and 80% relative humidity for 70°C and 50°C [47].

oxygen mole fraction decreases with decreasing pressure, and the proportion of water vapor mass increases. While at 1 bar total pressure, the partial pressure of oxygen is 0.16 for 80% relative humidity and 70°C, it decreases to 0.03 if the pressure is lowered to 0.3 bar. Reducing the temperature to 50°C means that at 0.3 bar, a partial pressure of 0.14 is still available, as shown on the right in **Figure** 7. Since the partial pressure of oxygen directly influences the performance of the cell, lower oxygen partial pressures will decrease the performance.

For the operation of a fuel cell system in aviation, this means that to maintain a constant relative humidity (rH) and constant O_2 concentration, the cathode air should always be compressed to operate the fuel cell at constant pressure. It is, therefore, unavoidable that in aircraft flying at high altitudes, the fuel cell system will be pressurised. However, in smaller aircraft with less than 1 MW propulsive power, as considered in the D-LIGHT project [46], which do not fly as high, non-pressurised systems could still be relevant. In this case, the pressure-dependent water management in the PEM fuel cell has a strong influence on performance and lifetime, and when operating at low pressures, the operating temperature should be lowered as well to enable operation with constant relative humidity while keeping O_2 content at reasonable levels.

The capability of the air to take on water and, therefore, to dry out the membrane can be quantified by the saturation deficit, which is defined in Eq. (8) as the difference between the saturation vapor pressure and the partial pressure of water.

$$SD = p_{sat} - p_{H_2O} \tag{8}$$

For a constant mass fraction of water in air and constant temperature, the saturation deficit increases with decreasing pressure, as shown in **Figure 8** for 150 g of water in 1 kg of dry air. The risk of drying out the membrane, therefore, increases. It can also be seen that for lower temperatures, the saturation deficit decreases.

Previous studies have also shown that temperature levels should be lowered for operation at lower pressures. Werner et al. [14] experimentally examined advisable operating temperatures, stoichiometry, and current for different pressure levels. Schroeter [48] calibrated an open-access model [49] using experimental data at 1 bar, 1.8 and 2.5 bar while treating the operating temperature as a fitting parameter. Both observed a lower temperature for operation at lower pressures. This supports

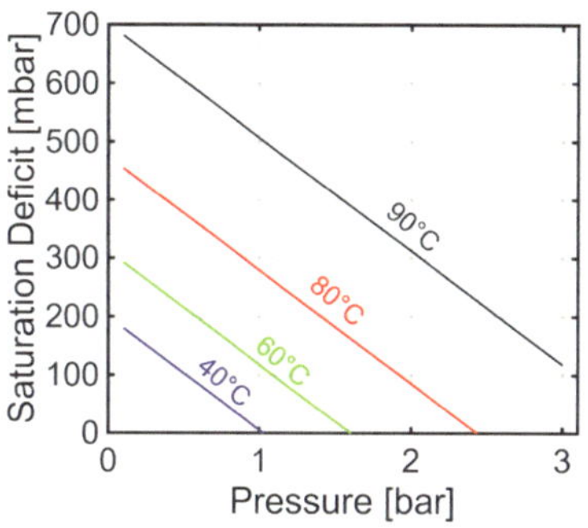

Figure 8.
Saturation deficit for different temperatures plotted against pressure for 150 g of water per kg of dry air.

the claim that operational temperature should be lowered for operation at lower pressures. If a fuel cell in an aircraft is operated without pressurisation or is pressurised with a constant pressure ratio, the operating pressure will change during operation according to the altitude. To ensure optimal operation of the fuel cell and reduce degradation, the temperature should be adjusted as well, resulting in the need for a suitable thermal management strategy.

4.3 HT-PEM technology to increase operational temperature

As fuel cell systems have an efficiency in the range of 60% [50], and other components (inverter, electric motor, etc.) produce heat as well, heat in the same order of magnitude as the propulsive power has to be rejected to the surroundings of the aircraft during operation. This will be especially challenging during take-off, when power requirements are highest and ambient air temperature may be high. At high altitudes, the lower temperatures will increase heat transfer, but low surrounding pressures reduce heat transfer to the surroundings [51].

Increasing the temperature level at which the fuel cell is operated promises to facilitate heat rejection to the surroundings, which is why the use of High temperature PEM fuel cells (HT-PEM) has lately gained considerable attention. HT technology promises to reach higher power densities at the system level once the technology reaches maturity, with 5–6 kW/kg expected by 2050, while LT PEM is only attributed the potential to reach 3 to 3.5 kW/kg by 2050 [10]. Both Airbus [52] and ZeroAvia [53] have announced their interest in developing the technology for aviation.

HT-PEM are operated at temperatures between 100 and 200°C [54], normally between 160 and 200°C [55]. The higher temperature facilitates the thermal management of the cell and the heat dissipation due to larger temperature differences. It also enables better use of the fuel cell's waste heat. While water/glycol mixtures are the predominant coolant for low-temperature fuel cell systems, other coolants are required for HT-PEM [56]. Although HT-PEM perform worse than low-temperature cells at low temperatures and low currents, they can reach higher current densities if operated correctly [57]. Another advantage of HT-PEM is that they do not rely on water for charge transport within the electrolyte [57]. This simplifies the system architecture, since water management with a humidifier for providing sufficient water to the cell is not required. Due to the elevated temperature, there is also no liquid water within the cells that can block active sites and lead to decreased performance or increased degradation. The higher temperatures above 160°C also lower the requirements for H_2 purity, since HT-PEM can tolerate up to 3% CO [54] in the fuel.

4.4 Integration of liquid hydrogen

The storage of sufficient hydrogen in an aircraft remains one of the most critical challenges to enable extended flight distances and higher propulsive power. Compressed gaseous hydrogen (GH_2) is unsuitable for large commercial aircraft because of its inherently low volumetric energy density [58]. The commercialisation of hydrogen-powered aircraft requires cryogenic storage in the form of liquid hydrogen (LH_2). **Figure 9** shows predicted flight distances and passenger numbers of hydrogen-based airplanes for different aircraft and aircraft concepts. It can be seen that all concepts for intermediate or long-distance aircraft base their considerations on liquid hydrogen. Only in the commuter and regional class is the use of GH_2 considered.

This integration of LH_2 requires the development of lightweight LH_2 tanks. For near-term evolutionary designs, a gravimetric index in the range of 0.2–0.35 (i.e., 20–35% of the fuel-system mass being usable hydrogen) is often assumed, whereas more aggressive composite-tank concepts report substantially higher GI values in specific tests [59].

The integration of LH_2 into the system also offers unique advantages, as the low temperatures and heat required for regasification can be integrated into the thermal management architecture, thereby supporting the cooling of fuel cells and power electronics, or even facilitating the integration of superconducting components, which increases the overall efficiency and specific power [60].

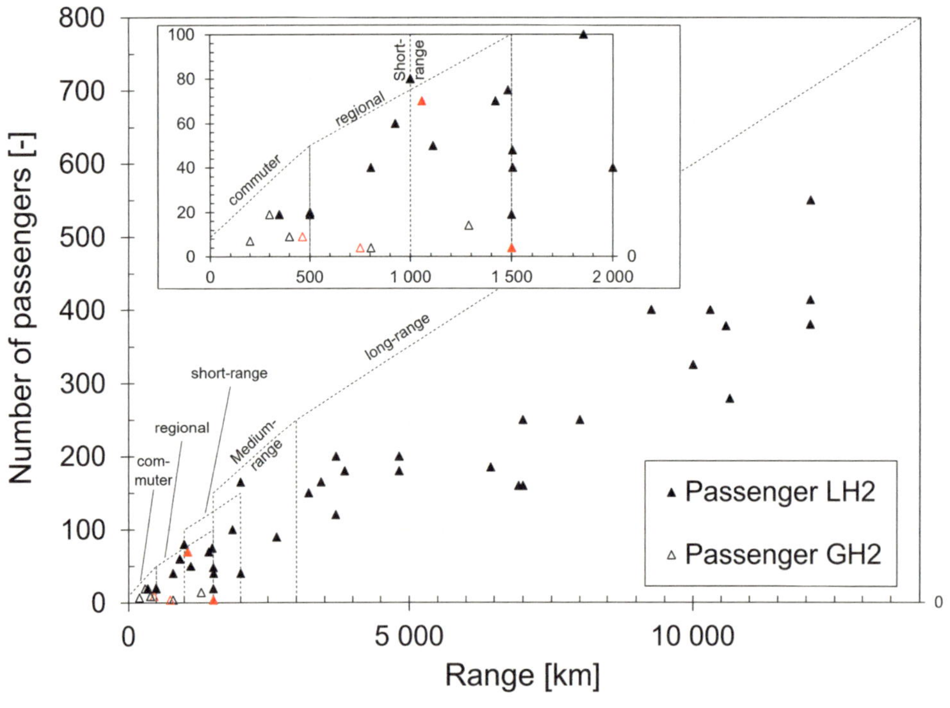

Figure 9.
Passenger numbers of conceptual hydrogen-based aircraft and existing demonstrators using liquid or gaseous hydrogen against flight range. Red symbols represent demonstrators that have already flown. Data is based on [61, 62].

Another central challenge concerns tank pressure management. The regulation of tank pressure via controlled vaporisation, venting strategies, or auxiliary pressurisation is essential to ensure stable hydrogen mass flow and to mitigate structural loads [63]. Material selection is equally critical: candidate materials must maintain integrity at cryogenic temperatures and resist hydrogen embrittlement, which can compromise metallic and composite structures under cyclic loading conditions [64]. Finally, safety considerations such as leakage, ignition risk, oxygen condensation on cold surfaces, and emergency venting strategies must be addressed in both design and operation [65]. Additionally, ground infrastructure for cryogenic refuelling and airport safety protocols must be developed [66]. In summary, the integration of liquid hydrogen into aviation constitutes a complex, multi-disciplinary systems challenge that extends far beyond tank design.

5. Conclusion

Fuel cells offer a high potential to decarbonise aviation. The use of fuel cells at high altitudes requires adaptations to fuel cell systems that were designed for operation on the ground. At high altitudes, the fuel cells have to be pressurised to compensate for the low ambient pressures, which would otherwise reduce the power output of the fuel cell. Pressurisation is achieved via an electric compressor. Depending on the target flight altitude, these compressors have to be designed to provide the required mass flow and desired outlet pressure; otherwise, the fuel cell may suffer due to air starvation, as was shown experimentally as well as through simulations. The varying ambient conditions also require a suitable control strategy for the compression system, which works reliably under different ambient conditions and avoids unstable situations, such as compressor surge. It was demonstrated that similarity values can be used to simplify the task and that surge control strategies prioritising mass flow can be successfully implemented.

Further optimisation potentials that should be addressed in the future are presented as well. A turbine in the exhaust of the cathode air can recuperate energy, increasing the overall output of a fuel cell system, while sophisticated water and temperature management in the fuel cell stack can increase performance at lower operational pressure, while extending the lifetime of the fuel cell stack. Integrating other technologies like HT-PEM fuel cells, as well as system integration with liquid hydrogen storage, also offers great potential for future hydrogen-powered aircraft.

Nomenclature

P_{el}	Electrical power
P_{is}	Power for isentropic compression
η	Efficiency
c_p	Specific heat capacity
κ	Isentropic exponent
T_{in}/T_{out}	Inlet/oulet temperature
T_{ref}/p_{ref}	Temperature/pressure at reference condition
p_{in}/p_{out}	Inlet/outlet pressure
Ψ	Pressure ratio

$\dot{m}$	Mass flow
$\dot{m}_{corr}$	Corrected massflow
λ	Cathode stoichiometry
p_{H_2O}	Partial pressure water in air
p_{sat}	Saturation pressure
rH	Relative humidity

Acknowledgments

The presented work was, in part, funded through the following projects:

- HighV (grant no: 20Y1701B), German Federal Ministry for Economic Affairs and Energy
- Go4Hy2 (Grant No: 03B10703A), German Federal Ministry for Digital and Transport
- Enable (Grant No: 20M1905D), German Federal Ministry for Economic Affairs and Energy
- Flüssigwasserstoffversorgung in einem Brennstoffzellenflugzeug (Grant Nr. 1.1412.2024.1), KSB Stiftung

Many thanks to the team at the Institute who supported or contributed to the presented work: Robin Fonk, Stefan Wazlawik, Marcel Haag, Jochen Mannes, and Fatih Türk. Also, thanks to Josef Kallo for contributing to the acquisition of some of the funding.

Conflict of Interest

The authors declare no conflicts of interest.

Author details

Caroline Willich*, Tobias Graf, Daniel Frank, Roland Ademi, Samara Brandão de Almeida and Christiane Bauer
Ulm University, Institute for Energy Conversion and Storage, Ulm, Germany

*Address all correspondence to: caroline.willich@uni-ulm.de

References

[1] Fuel Cells and Hydrogen 2 Joint Undertaking. Hydrogen-powered aviation: A fact based study of hydrogen technology, economics, and climate impact by 2050. [Internet]. LU: Publications Office; 2020. Available from: https://data.europa.eu/doi/10.2843/471510 [Accessed: 2024-July-25]

[2] Rho Motion. Market and Technology Assessment of Grid-Scale Energy Storage required to Deliver Net Zero and the Implications for Battery Research in the UK. Faraday Institution; 2023. Available from: https://www.faraday.ac.uk/wp-content/uploads/2023/09/20230908_Rho_Motion_Faraday_Institution_UK_BESS_Report_Final.pdf [Accessed: 2026-January-03]

[3] Koehler T. A green machine, Boeing makes history with flights of Fuel Cell Demonstrator Airplane. [Internet]. Available from: https://cdn.archilovers.com/projects/43ae05e9-c1be-4065-a15c-f3038bcb9dec.pdf [Accessed: 2024-September-24]

[4] H2FLY GmbH. H2FLY [Internet]. Available from: https://www.h2fly.de/ [Accessed: 2022-September-29]

[5] Doll S. H2FLY completes world's first piloted flight of a liquid hydrogen powered electric aircraft. [Internet]. Available from: https://electrek.co/2023/09/07/h2fly-completes-worlds-first-piloted-flight-liquid-hydrogen-electric-aircraft-video/ [Accessed: 2024-September-26]

[6] Reichmann K. ZeroAvia Completes First Hydrogen-Powered Flight. Aviation Today. 2020. Available form: https://www.aviationtoday.com/2020/09/29/zeroavia-completes-first-hydrogen-electric-turboprop-flight/ [Accessed 2026-January-03]

[7] Lundbladh A, Grönstedt T, Genrup M, Platell P High Power Density Work Extraction from Turbofan Exhaust Heat. In: Conference of the International Society for Air Breathing Engines. 2015.

[8] the Superstack M-The Fuel Cell Power Generation System for ZeroAvia's 600kW Powertrains. [Internet] ZeroAvia. Available from: https://zeroavia.com/blogs/meet-the-superstack/ [Accessed: 2025-July-23]

[9] Padgett E, Kleen G. Automotive Fuel Cell Targets and Status. USA: Department of Energy; 2020

[10] Bhatti W, Wei W, Doyle F, Llambrich J, Webber H, Town N. FUEL CELLS Roadmap Report [Internet]. Aerospace Technology Institute; 2022. Available from: https://www.ati.org.uk/wp-content/uploads/2022/03/FZO-PPN-COM-0033-Fuel-Cells-Roadmap-Report.pdf. [Accessed 2026- February-03]

[11] International Standard Atmosphere (ISA) Table. Performance of the Jet Transport Airplane: Analysis Methods, Flight Operations and Regulations [Internet]. Wiley; 2017. p. 583–590. Available from: https://onlinelibrary.wiley.com/doi/10.1002/9781118534786.app1 [Accessed: 2025-July-22]

[12] Morris E What is Cruising Altitude? [Internet]. Sheffield School of Aeronautics. 2022. Available from: https://www.sheffield.com/2022/what-is-cruising-altitude.html [Accessed: 2025-July-22]

[13] Mildrexler DJ, Zhao M, Running SW. Where are the hottest spots on Earth? EoS Transactions. 2006;**87**(43):461–467

[14] Werner C, Gores F, Busemeyer L, Kallo J, Heitmann S, Griebenow M. Characteristics of PEMFC operation in ambient- and low-pressure environment considering the fuel cell humidification. CEAS Aeronautical Journal. 2015;**6**(2):229–243

[15] Henke M Pressurised solid oxide fuel cells: From electrode electrochemistry to hybrid power plant system integration [Internet]. Universität Stuttgart; 2015. Available from: https://elib.uni-stuttgart.de/handle/11682/2409 [Accessed: 2024-September-17]

[16] Pratt J, Brouwer J, Samuelsen G Experimental Performance of an Air-Breathing PEM Fuel Cell at High Altitude Conditions. In: 43rd AIAA Aerospace Sciences Meeting and Exhibit [Internet]. Reno, Nevada: American Institute of Aeronautics and Astronautics; 2005. Available from: https://arc.aiaa.org/doi/10.2514/6.2005-953 [Accessed: 2025-August-20]

[17] Atkins P. Physical Chemistry. 5th ed. Oxford, Melbourne, Tokyo: Oxford University press; 1994

[18] Stühmeier BM, Pietsch MR, Schwämmlein JN, Gasteiger HA. Pressure and Temperature Dependence of the Hydrogen Oxidation and Evolution Reaction Kinetics on Pt Electrocatalysts via PEMFC-based Hydrogen-Pump Measurements. Journal of the Electrochemical Society. 2021;**168**(6):064516

[19] Berning T. The dew point temperature as a criterion for optimizing the operating conditions of proton exchange membrane fuel cells. International Journal of Hydrogen Energy. 2012;**37**(13):10265–10275

[20] Sigwadi R, Dhlamini MS, Mokrani T, N̂emavhola F, Nonjola PF, Msomi PF. The proton conductivity and mechanical properties of Nafion®/ ZrP nanocomposite membrane. Heliyon. 2019;**5**(8):e02240

[21] Willich C, Frank D, Graf T, Wazlawik S, Brandao S, Bauer C. High-Altitude Operation of a Commercial 100 kW PEM Fuel Cell System. Energies. 2024;**17**(24):6309. Available from: https://www.mdpi.com/1996-1073/17/24/6309

[22] Jung SH, Kim SL, Kim MS, Park Y, Lim TW. Experimental study of gas humidification with injectors for automotive PEM fuel cell systems. Journal of Power Sources. 2007;**170**(2):324–333

[23] Chen D, Li W, Peng H. An experimental study and model validation of a membrane humidifier for PEM fuel cell humidification control. Journal of Power Sources. 2008;**180**(1):461–467

[24] Chang Y, Qin Y, Yin Y, Zhang J, Li X. Humidification strategy for polymer electrolyte membrane fuel cells–A review. Applied Energy. 2018;**230**:643–662

[25] Willich C A direct fuel-cell-based hybrid system for airborne applications [Internet]. Universität Ulm; 2024. Available from: https://oparu.uni-ulm.de//handle/123456789/54870 [Accessed: 2025-January-08]

[26] Filsinger D, Kuwata G, Ikeya N. Tailored Centrifugal Turbomachinery for Electric Fuel Cell Turbocharger.

International Journal of Rotating Machinery. 2021;**2021**:1–14

[27] Lück S, Wittmann T, Göing J, Bode C, Friedrichs J. Impact of Condensation on the System Performance of a Fuel Cell Turbocharger. Machines. 2022;**10**(1):59

[28] Antivachis M, Dietz F, Zwyssig C, Bortis D, Kolar JW. Novel High-Speed Turbo Compressor With Integrated Inverter for Fuel Cell Air Supply. Frontiers of Mechanical Engineering. 2021;**6**:612301

[29] Yumiya H, Kizaki M, Asai H. Toyota Fuel Cell System (TFCS). Wevj. 2015;7(1):85–92

[30] Toyota Industries Develops New FC Air Compressor and Hydrogen Circulation Pump, Now Installed in the New Toyota Mirai FCEV. Toyota Industries Corporation [Internet]. Available from: https://www.toyota-industries.com/news/2020/12/10/005056/index.html [Accessed: 2024-October-28]

[31] Radica G, Tolj I, Lototskyy MV, Pasupathi S. Air Mass Flow and Pressure Optimization of a PEM Fuel Cell Hybrid System for a Forklift Application. Energies. 2023;**17**(1):120

[32] ElringKlinger AG. Elektromobilität. PEM-Brennstoffzellen: hocheffizient, umweltgerecht, zukunftsfähig. ElringKlinger; 2019. Available from: http://www.elringklinger.de/ [Accessed: 2019-June-17]

[33] Eckert P, Hemmer S, Bouwman R. Results from demonstration of the range extender [Internet]. 2019-Dec Report No: D 6.3. Available from: https://ec.europa.eu/research/participants/documents/downloadPublic?documentIds=080166e5ca78dfd8&appId=PPGMS [Accessed: 2025-September-19]

[34] Schröter J, Graf T, Frank D, Bauer C, Kallo J, Willich C. Influence of pressure losses on compressor performance in a pressurized fuel cell air supply system for airplane applications. International Journal of Hydrogen Energy. 2021;**46**(40):21151–21159

[35] Bruckner RJ Windage Power Loss in Gas Foil Bearings and the Rotor-Stator Clearance of High Speed Generators Operating in High Pressure Environments. In: Volume 5: Microturbines and Small Turbomachinery; Oil and Gas Applications [Internet]. Orlando, Florida, USA: ASMEDC; 2009. p. 263–270. Available from: https://asmedigitalcollection.asme.org/GT/proceedings/GT2009/48869/263/344234 [Accessed: 2025-May-20]

[36] Brandao de Almeida SG, Willich C, Bauer C. Modelling an Air Supply System for Fuel Cell Applications in Aviation, AIAA SCITECH 2024 Forum. Orlando, FL, USA: 2024. DOI: 10.2514/6.2024-1533

[37] Frank D, Schröter J, Bauer C, Willich C. Development of a control strategy for an air supply module for pressurizing fuel cells in airborne applications. International Journal of Hydrogen Energy. 2024;**51**:1374–1381

[38] Frank D, Wazlawik S, Brandao de Almeida S, Bauer C, Willich C. Model-Based Development of Control Strategies for a Fuel Cell Air Supply System and Experimental Verification at Flight-Relevant Conditions, PRiME 2024. Honululu, Hawaii: 2024

[39] Gravdahl JT, Egeland O. Compressor Surge and Rotating Stall. In: Grimble MJ, Johnson MA, editors. Advances in Industrial Control. [Internet]. London: Springer London; 1999. DOI: 10.1007/978-1-4471-0827-6 [Accessed: 2026-February-03]

[40] Kadyk T, Winnefeld C, Hanke-Rauschenbach R, Krewer U. Analysis and Design of Fuel Cell Systems for Aviation. Energies. 2018;**11** (2):375

[41] Friedrich KA, Kallo J, Schirmer J, Schmitthals G. Fuel Cell Systems for Aircraft Application. ECS Transactions. 2009;**25**(1):193–202

[42] Zhang F, Maddy J Investigation of the Challenges and Issues of Hydrogen and Hydrogen Fuel Cell Applications in Aviation [Internet]. 2021. Available from: https://www.techrxiv.org/articles/preprint/Investigation_of_the_Challenges_and_Issues_of_Hydrogen_and_Hydrogen_Fuel_Cell_Applications_in_Aviation/14958057 [Accessed: 2025-September-26]

[43] Lück S, Göing J, Wittmann T, Mimic D, Friedrichs J. Towards design-and operating-point selection for fuel cell cathode air-supply systems in aviation. Jgpp. 2024;**15**(2):76–84

[44] Larminie J, Dicks A. Fuel Cell Systems Explained. UK: John Wiley & Sons Ltd; 2003. DOI: 10.1002/9781118878330

[45] World Meteorological Organization (WMO). Guide to Instruments and Methods of Observation (WMO-No. 8), Volume I – Measurement of Meteorological Variables. [Internet]. World Meteorological Organization (WMO); 2024. Available from: https://library.wmo.int/idurl/4/68695 [Accessed: 2025-September-26]

[46] DLR Technologien für Kleinflugzeuge. D-LIGHT [Internet]. Available from: https://www.dlr.de/de/kf/forschung-transfer/projekte/d-light [Accessed: 2025-September-23]

[47] Graf T, Frank D, Brandao S, Wazlawik S, Kley C, Bauer C, et al. Commercial PEM Fuel Cell System in Aviation 19th Ulm Electrochemical Days (UECT). Ulm, Germany: June 2025

[48] Schröter J Fuel cell air supply for hydrogen electric propulsion systems in aircraft applications. 2023-Jun-13; Available from: https://oparu.uni-ulm.de/items/14a54952-7951-488c-ace1-d369ea64d0a2 [Accessed: 2024-April-02]

[49] Vetter R, Schumacher JO. Free open reference implementation of a two-phase PEM fuel cell model. Computer Physics Communications. 2019;**234**:223–234

[50] Boukoberine MN, Zhou Z, Benbouzid M. A critical review on unmanned aerial vehicles power supply and energy management: Solutions, strategies, and prospects. Applied Energy. 2019;**255**:113823

[51] Liu W, Wang J, Li Y, Zhu Z, Qie D, Ding L. Natural convection heat transfer at reduced pressures. Experimental Heat Transfer. 2019;**32**(1):14–24

[52] Advent Technologies. Revolutionizing Aviation: Propelling Towards Zero Emissions with HT-PEM Fuel Cell Technology. [Internet]. Advent Technologies. 2023. Available from: https://advent.energy/2023/11/24/revolutionizing-aviation-propelling-

towards-zero-emissions-with-ht-pem-fuel-cell/ [Accessed: 2024-April-30]

[53] Joint News Release - BASF Environmental Catalyst and Metal Solutions (ECMS) and ZeroAvia partner to advance high temperature proton exchange membrane (HT-PEM) fuel cells for hydrogen-fueled aviation [Internet]. Available from: https://www.basf.com/global/en/media/news-releases/2023/09/p-23-319.html [Accessed: 2024-April-30]

[54] Chandan A, Hattenberger M, El-kharouf A, Du S, Dhir A, Self V, et al High temperature (HT) polymer electrolyte membrane fuel cells (PEMFC) – A review. Journal of Power Sources. 2013;**231**:264–278

[55] Waller MG, Walluk MR, Trabold TA. Performance of high temperature PEM fuel cell materials. Part 1: Effects of temperature, pressure and anode dilution. International Journal of Hydrogen Energy. 2016;**41**(4):2944–2954

[56] Nasri M, Dickinson D Thermal management of fuel cell-driven vehicles using HT-PEM and hydrogen storage. In: 2014 Ninth International Conference on Ecological Vehicles and Renewable Energies (EVER). Monte-Carlo: IEEE; 2014. p. 1–6. Available from: https://ieeexplore.ieee.org/document/6844107/ [Accessed: 2025-September-26]

[57] Authayanun S, Mamlouk M, Scott K, Arpornwichanop A. Comparison of high-temperature and low temperature polymer electrolyte membrane fuel cell systems with glycerol reforming process for stationary applications. Applied Energy. 2013;**109**:192–201

[58] Andrews J, Shabani B. Re-envisioning the role of hydrogen in a sustainable energy economy. International Journal of Hydrogen Energy. 2012;**37**(2):1184–1203

[59] Mukhopadhaya J, Rutherford D. Performance Analysis of Evolutionary Hydrogen-powered Aircraft. [Internet] International Council on Clean Transportation; 2022: Available from: https://theicct.org/wp-content/uploads/2022/01/LH2-aircraft-white-paper-A4-v4.pdf [Accessed: 2025-September-26]

[60] Nam GD, Vuong LD, Sung HJ, Lee SJ, Park M. Conceptual Design of an Aviation Propulsion System Using Hydrogen Fuel Cell and Superconducting Motor. IEEE Transactions on Applied Superconductivity. 2021;**31**(5):1–7

[61] Tiwari S, Pekris MJ, Doherty JJ. A review of liquid hydrogen aircraft and propulsion technologies. International Journal of Hydrogen Energy. 2024;**57**:1174–1196

[62] Adler EJ, Martins JRRA. Hydrogen-powered aircraft: Fundamental concepts, key technologies, and environmental impacts. Progress in Aerospace Sciences. 2023;**141**:100922

[63] Huete J, Pilidis P. Parametric study on tank integration for hydrogen civil aviation propulsion. International Journal of Hydrogen Energy. 2021;**46**(74):37049–37062

[64] Ilyushechkin A, Schoeman L, Carter L, Hla SS. Material Challenges and Hydrogen Embrittlement Assessment for Hydrogen Utilisation in Industrial Scale. Hydrogen. 2023;**4**(3):599–619

[65] Anekwe IMS, Mustapha SI, Akpasi SO, Tetteh EK, Joel AS, Isa YM. The hydrogen challenge: Addressing

storage, safety, and environmental concerns in hydrogen economy. International Journal of Hydrogen Energy. 2025;**167**:150952

[66] Hoelzen J, Koenemann L, Kistner L, Schenke F, Bensmann A, Hanke-Rauschenbach R. H2-powered aviation–Design and economics of green LH2 supply for airports. Energy Conversion and Management: X. 2023;**20**:100442

Chapter 5

Post-Pandemic Airport Capacity Optimization for Sustainable Aviation Growth

Woojin Choi

Abstract

This chapter examines the critical imperative of sustainable airport capacity expansion and management in an era of unprecedented air traffic growth and heightened uncertainty. Despite a robust post-pandemic recovery that has pushed global passenger traffic beyond 2019 levels, the industry faces significant challenges from geopolitical tensions, supply chain disruptions, and capacity shortages. The traditional model of reactive, physical expansion is no longer sufficient. Instead, a shift toward a flexible, data-driven approach is necessary to reconcile growth with environmental stewardship and operational resilience. We discuss five core themes shaping the future of airports: demand uncertainty, sustainability, smart technologies, flexibility, and their evolution into integrated urban platforms. The chapter presents case studies, including Western Sydney Airport, to illustrate how airports are leveraging sustainable building practices and advanced planning models to safeguard future expansion and innovation needs, mitigate environmental impacts, and enhance resilience. Additionally, we explore the role of digital transformation, driven by artificial intelligence (AI) and data analytics, in optimizing operations, improving passenger experience, and enabling energy efficiency. The chapter concludes by outlining a strategic framework for future-proof airport development, emphasizing the importance of collaborative data platforms, modular infrastructure, and stakeholder engagement to navigate a volatile and rapidly evolving landscape.

Keywords: airport capacity development, capacity optimization modeling, airport post-pandemic strategy, sustainable airport development, innovative airport technology, air traffic demand uncertainty

1. Introduction

Airports are undergoing a profound transformation, evolving from mere transport hubs into complex, interconnected ecosystems that significantly influence global and urban economies. Beyond their primary function of facilitating air travel, airports now serve as catalysts for regional development, creating employment opportunities and driving economic activity through a diverse range of aeronautical and nonaeronautical businesses. Consequently, effective airport capacity planning and operations must extend beyond infrastructure considerations to encompass the broader socioeconomic and environmental contexts, ensuring harmonious and sustainable development.

According to the Airports Council International (ACI) World Airport Traffic Dataset, global passenger traffic reached an all-time high of over 9.4 billion passengers in 2024, marking an 8.4% increase from 2023 and a 2.7% rise above pre-pandemic levels in 2019 [1]. This milestone signifies a strong recovery for the air travel industry. ACI forecasts continued growth, with passenger traffic expected to reach 9.9 billion in 2025, representing a 4.8% year-over-year increase [2].

However, this growth trajectory is not without challenges. The post-pandemic environment is characterized by significant uncertainties stemming from economic and geopolitical landscapes. Ongoing supply chain disruptions and production delays from aircraft manufacturers, coupled with geopolitical tensions, pose risks to future growth by potentially altering flight routes and increasing operational costs. Additionally, the threat of new tariffs could disrupt global trade, which would, in turn, affect international travel demand and future airport capacity requirements. Although passenger demand is projected to remain robust, the pace of capacity expansion may not keep pace with this growth. This mismatch necessitates a shift from a reactive, short-term approach to a more sustainable, structural, and long-term strategy for airport capacity expansion and operational resilience.

Simultaneously, technological advancements, particularly in data science and artificial intelligence (AI), are reshaping travel patterns and requiring a redefinition of airport processes. These innovations offer the potential to create a high-quality, seamless, and personalized travel experience. Airports are, therefore, becoming "experience centers" and destinations, functioning as vital centers for global business and as integral components of urban life.

To address these multifaceted challenges, sustainable and smart airport capacity expansion requires a flexible, data-driven approach and close coordination among the various stakeholders. This strategy involves integrating advanced technologies, optimizing existing infrastructure, and prioritizing environmental sustainability to ensure long-term resilience and efficiency. A key objective of sustainable airport capacity planning is to reconcile infrastructure growth with environmental stewardship, thereby mitigating impacts such as decarbonization, noise pollution, and territorial impacts, while leveraging technological innovation and energy efficiency.

This chapter will discuss both the key challenges and opportunities for airport development and management, with a strong focus on the following five core themes:

- **Demand uncertainty**: Air traffic is inherently volatile due to economic fluctuations, global events, and unforeseen crises. Airports cannot eliminate this unpredictability but can build resilience and adaptability into their planning and operations.

- **Sustainability**: This principle integrates the social benefits of air transport with environmental protection at both local and global levels. Airports must proactively minimize their environmental footprint by reducing carbon emissions, noise pollution, and waste generation.

DOI: http://dx.doi.org/10.5772/intechopen.1013181

- **Smart technologies**: The integration of technologies such as AI, machine learning, and automation is crucial for optimizing operational efficiency, enhancing the passenger experience, and improving resource allocation.

- **Flexibility and adaptability**: Airport infrastructure must be designed to accommodate fluctuating demand and evolving technologies, while maintaining a high level of service to meet changing customer expectations.

- **Transitioning into an integrated urban platform**: Airports are increasingly evolving beyond their traditional role as transport hubs. This trend involves integrating aviation with commercial and community functions, positioning airports as interconnected urban ecosystems that support diverse services, including retail, hospitality, logistics, and mobility.

2. The sustainability imperative in airport development and capacity expansion

Airports are often significant sources of environmental impact, negatively affecting neighboring communities and natural ecosystems. The construction and operation of an airport can disrupt land and waterways, leading to habitat loss and a continuous impact on the surrounding area through noise pollution, greenhouse gas emissions, and other airborne pollutants.

Enhancing operational efficiency is a powerful strategy to reduce emissions, conserve resources, and lower costs. The electrification of ground infrastructure and the optimization of airside operations can contribute to decarbonization and improve air quality while simultaneously boosting profitability. For instance, the efficient use of runways and taxiways can reduce emissions and increase operational efficiency, generating both social and economic benefits. As the industry moves toward net-zero emissions, airports must continue to invest in renewable energy, energy-efficient infrastructure, and support the adoption of sustainable aviation fuels (SAFs) and the electrification of ground operations.

Furthermore, airports should focus on resource conservation, waste reduction, and the implementation of responsible water and energy management programs. Embracing circular economy principles – reusing materials and minimizing waste through recycling – is essential and can further enhance sustainability efforts.

The pursuit of green building certifications, such as LEED (US), BREEAM (UK), Estidama (UAE), and Green Star (Australia), has become an essential requirement in sustainable airport planning, design, renovation, and expansion projects.

The importance of sustainable airport capacity expansion can be summarized as follows:

- **Adaptation to fluctuating demand**: Sustainable expansion allows airports to adjust to the shifting patterns of post-pandemic air travel, ensuring they can handle both peak and off-peak periods effectively.

- **Support for economic growth**: By enabling the efficient movement of people and goods, adequate capacity is vital for supporting tourism, business travel, and overall economic activity.

- **Integration of technological advancements**: Sustainable growth provides an opportunity to embed new technologies, such as automated baggage handling and biometric identification, to improve efficiency, security, and the passenger experience.

- **Environmental mitigation**: Implementing environmentally-friendly practices is a core component of sustainable growth, aligning airports with global climate goals and addressing public concerns about climate change.

- **Catalysis for urban development**: Airports are crucial for attracting businesses and creating jobs. Sustainable expansion ensures that this growth is aligned with the needs of urban populations and contributes positively to regional development.

- **Enhanced resilience**: In the face of increasing geopolitical instability and other disruptions, sustainable expansion helps airports become more resilient, ensuring continued operations during unforeseen events.

Recent airport cases highlight the growing trend of integrating sustainability into strategic airport development and management:

- Istanbul's new airport has earned a LEED Gold certification.

- Hamburg Airport became the first major German airport to achieve CO2-neutral operations.

- San Francisco International Airport's Terminal 1 is the first airport terminal globally to earn LEED Platinum.

- Perth Airport has committed to becoming carbon-neutral by 2030.

- Shell is offering sustainable aviation fuel (SAF) at Singapore Changi Airport.

- Christchurch Airport has dedicated 400 hectares to renewable energy generation.

- Vehicles at Edinburgh Airport are powered by vegetable oil.

- The FAA and U.S. airports have launched the Airport Climate Challenge to achieve net-zero emissions by 2050.

- Milan Airport and Airbus have signed an MoU for hydrogen distribution, with the first hydrogen-powered aircraft expected by 2035.

- Louisville Muhammad Ali International Airport is utilizing geothermal energy for its terminal.

- Austin-Bergstrom Airport has a new 1.8-megawatt solar panel array.

- Doha's Hamad Airport has partnered with the Ministry of Environment to improve waste management.

The COVID-19 pandemic further underscored the importance of resilience and adaptability as central components of airport sustainability. Airports are now prioritizing flexible infrastructure, diversified revenue streams, digital transformation, and robust health and safety measures. Planning requirements increasingly emphasize scalable capacity, enhanced cargo and logistics, and crisis preparedness, ensuring airports can withstand future disruptions while meeting climate goals and community expectations.

Sustainability opportunities in airport planning and development can be viewed through the lens of two major industries:

- **Aviation industry:** A significant opportunity lies in incorporating SAF production and storage, electrification, process automation, and digital technologies into airport planning. By facilitating this approach, airports can directly reduce lifecycle carbon emissions, support airline decarbonization targets, and position themselves as leaders in sustainable air transport.

- **Building industry:** Airports can leverage sustainable building design and operations to reduce emissions and operational costs. This includes adopting net-zero energy terminals through on-site renewable energy generation (e.g., solar PV, geothermal), advanced building management systems, and a circular resource economy (e.g., water recycling, waste-to-energy).

Figure 1 provides a framework for sustainable airport infrastructure development and operations from a lifecycle perspective.

2.1 Case study: Western Sydney Airport development project

The Western Sydney Airport project has embedded sustainability as a core pillar of its development, aligning with Australia's national climate commitments. The project focuses on net-zero readiness, climate resilience, and social value creation across design, construction, and operations. **Figure 2** illustrates key use cases implemented in the project.

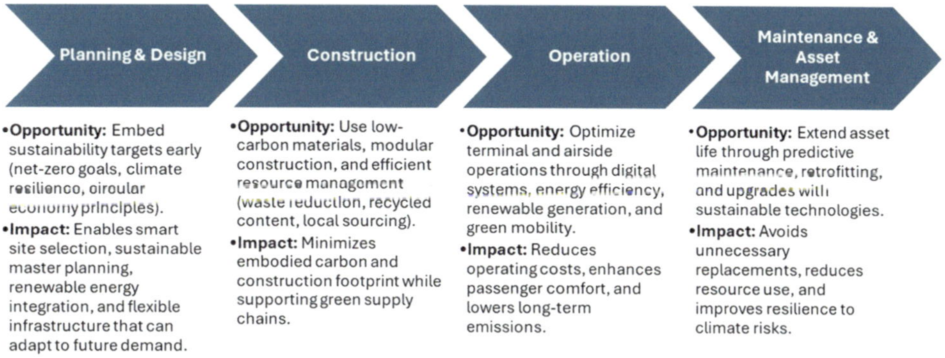

Figure 1.
Sustainable airport infrastructure development and operation framework.

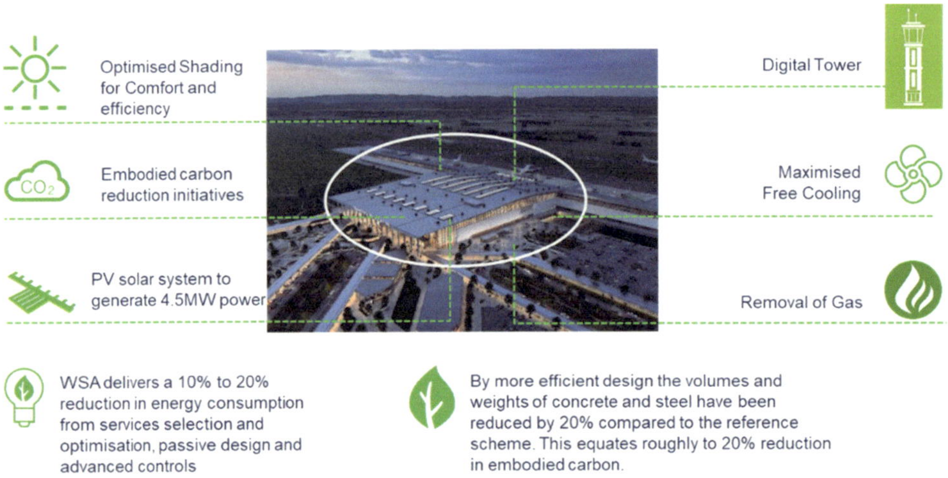

Figure 2.
Key use cases implemented within the Western Sydney Airport development project. Source: WSA Sustainability Plan [3].

- **Terminal design**: The terminal is designed to minimize embodied and operational carbon. This includes energy-efficient design, material selection, on-site solar generation, and electrification of ground support equipment and commercial operations to support the net-zero initiative. Boarding gates and baggage reclaim areas are designed for swing operations, enabling flexible use of facilities to efficiently accommodate varying peak traffic demands.

- **Energy and carbon management**: The project integrates renewable energy and smart building systems, with a long-term pathway toward net-zero operations by 2050.

- **Sustainable infrastructure and materials**: The use of low-carbon and recycled materials, along with circular economy strategies, minimizes construction waste. The modular design ensures future adaptability.

- **Water and biodiversity**: The project incorporates stormwater harvesting, Water-sensitive urban design (WSUD), and biodiversity protection through green corridors and local landscaping.

- **Mobility and connectivity**: A key feature is the seamless integration with the Sydney Metro, reducing reliance on private vehicles. The design also includes provisions for future electric and hydrogen vehicle charging infrastructure.

Additionally, the implementation of a digital tower is a significant sustainability initiative, enabling off-site operations and automated air traffic control. Various safeguarding solutions have been integrated into the design to ensure long-term capacity and adaptability, including modular structural design, service corridors for future utilities, and provisions for future technologies such as Vertical Take-off and Landing (VTOL) aircraft and hydrogen-powered vehicles.

2.2 Case study: Royal Schiphol Group's circular vision

Royal Schiphol Group aims to be a zero-waste airport by 2030 and a fully circular, energy-positive airport by 2050. This ambition is anchored in its Vision 2050, structured around four sustainability pillars:

- Energy positive – generating more renewable energy than the airport consumes.
- Circular economy – designing out waste and preserving materials through reuse and recycling.
- Sustainable aviation – supporting emissions reductions across the aviation value chain.
- Community engagement – ensuring operations enhance societal well-being.

As part of this strategy, Schiphol has adopted urban mining, setting up an on-site concrete recycling facility that reprocesses demolition rubble – saving transport emissions and supporting internal construction demand. An EME-based material marketplace helps catalog and reuse construction materials across projects.

2.2.1 Checkpoint 90: A model of circular design

Checkpoint 90 at Schiphol Airport exemplifies the practical implementation of the group's circular vision [4]. Key features include:

- **Urban mining in action:** Built primarily using salvaged materials – steel, doors, lighting, ceilings, and kitchen components – from three office buildings and three cargo structures previously demolished on-site.
- **High circularity score:** Achieves a Building Circularity Index (BCI) of 68–70%, nearing the design target and the highest possible score of 80%.
- **Energy-neutral and gas-free:** The building is fully off the gas grid, energy-neutral (Energy Label A++++), and features bifacial solar panels as a translucent, waterproof roof.
- **Demountable and reusable design:** Engineered for disassembly; materials and design principles serve as a toolbox for future checkpoints.
- **Broader circular impact:** Excess materials are repurposed; for example, roofing is turned into asphalt for Rotterdam Hague Airport, and unused frames are reused elsewhere on-site.

3. Optimizing airport capacity development under demand uncertainty

Delivering reliable performance in the complex airport environment is a significant challenge due to factors such as varying demand, diverse demographics, and operational complexity. The expansion of airport capacity in metropolitan areas, in particular, is complicated by stakeholder dynamics and substantial investment requirements. Available solutions include both multi-airport development and single-airport expansion. However, financial constraints and demand uncertainty can lead to protracted planning processes, hindering sustainable growth.

In the context of recent globalization, a cost-competitive business environment, and unforeseen crises like the COVID-19 pandemic, national or metropolitan-level capacity planning requires optimal solutions that account for future demand fluctuations and urban expansion. Historical trends show consistent growth in air travel, even after major crises. As shown in **Figure 3**, the post-pandemic recovery highlights the industry's resilience, despite new uncertainties arising from geopolitical events.

While global passenger traffic has surpassed 2019 levels, performance gaps across individual airport markets are becoming the "new normal." Major metropolitan airports are experiencing increasing congestion, leading to crowded terminals and longer waiting times. **Figure 4** illustrates that a significant number of the world's top

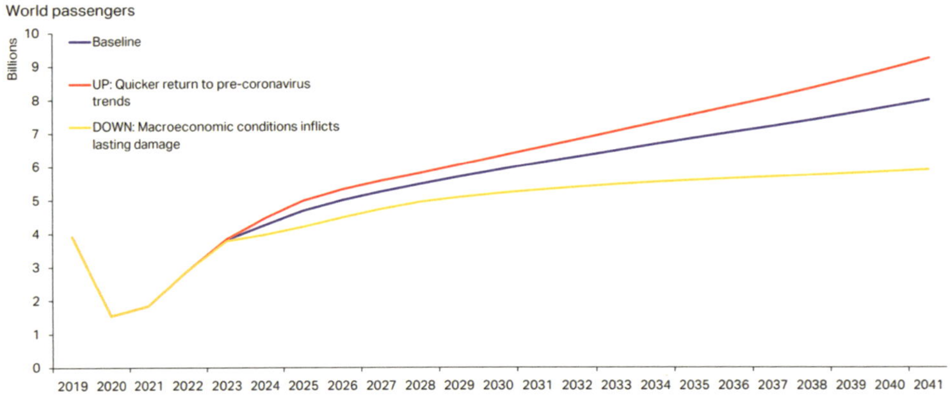

Figure 3.
Post-pandemic air traffic demand forecast scenarios. Source: Tourism Economics/IATA Sustainability and Economics [2].

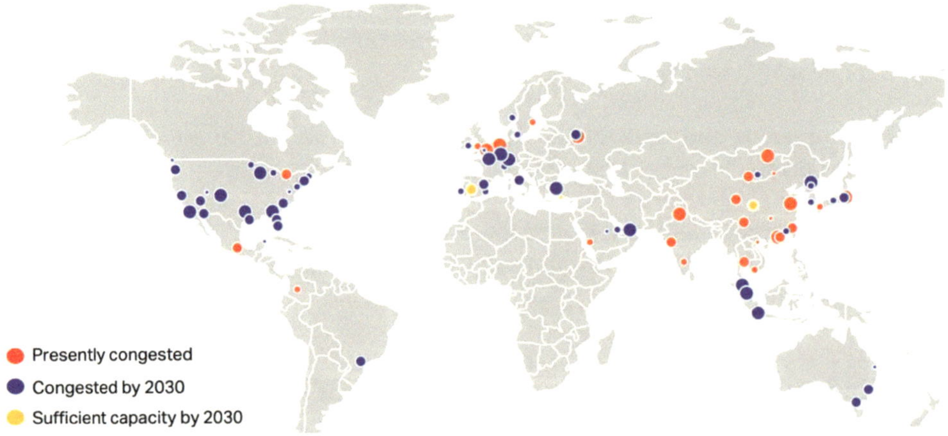

Figure 4.
Status of the top 100 busiest airport congestions. Source: IATA Global Airport and Passenger Symposium [5].

100 airports are already congested, with more projected to reach critical capacity within the next five years.

Airport congestion has negative consequences for passenger services, turnaround processes, socio-economic costs, and operational safety. Conversely, providing excessive capacity can result in unnecessary operational costs and energy consumption from underutilized infrastructure.

Given the complexity of these issues, a robust, data-driven decision-making framework is essential to optimize airport capacity expansion. A combined approach – integrating linear programming (LP) with stochastic simulation or advanced data analytics – can provide a strong foundation for building an effective optimization model that supports stakeholder decision-making.

Since airport capacity planning involves a wide range of variables, interdependencies, and nonlinear relationships, the use of mixed-integer nonlinear programming (MINLP) is particularly well-suited. This methodology enables the modeling of complex constraints and discrete choices, making it highly effective for developing realistic and actionable capacity expansion strategies.

The four primary solutions for addressing airport capacity challenges in metropolitan areas are illustrated in **Figure 5**, along with global example cities and summarized below:

- **Solution 1:** Expand the capacity of an existing airport until it reaches its maximum operational limit.

- **Solution 2:** Develop a new airport, close the existing one, and transfer all traffic to the new facility.

- **Solution 3:** Develop a new airport to operate alongside the existing airport(s), creating a multi-airport system.

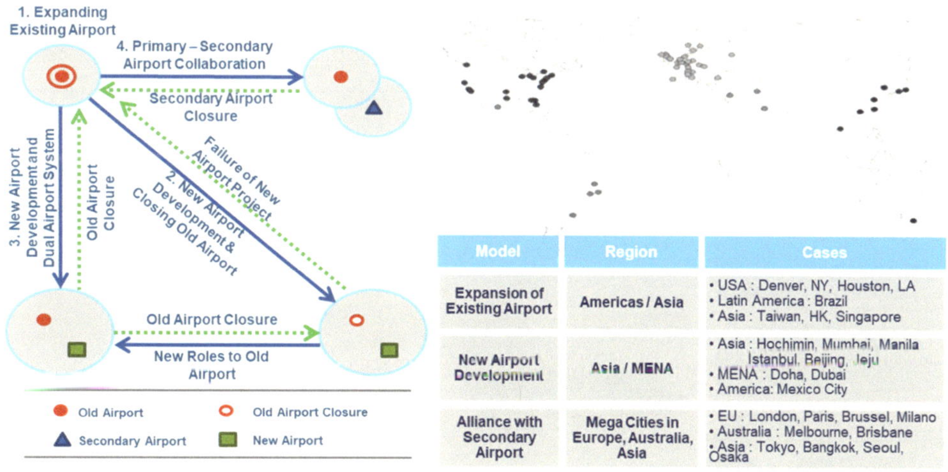

Figure 5.
Major solutions for airport capacity expansion in metropolitan regions. Conceptualized airport capacity expansion scenarios sketched by the author. Solutions 1 and 2 are regarded as a single-airport system, whereas Solutions 3 and 4 fall under a multi-airport system. From an empirical view, Solutions 2, 3, and 4 are pursued after the practice of Solution 1.

- **Solution 4:** Modernize or repurpose underutilized general aviation or military airports into secondary commercial airports to support the primary hub.

Today's airport capacity planning has shifted from relying solely on historical data/trends to a more dynamic approach. As shown in **Figure 6**, current planning aims to:

- Understand passengers' needs and identify potential bottlenecks to ensure airport capacity is effectively aligned with traffic demand profiles.

- Leverage technology, data, and operational procedures to create new passenger experiences, as well as enhance airport capacity efficiency.

- Prioritize community and environmental commitments to support sustainable and socially responsible airport development.

Key requirements for sustainable airport capacity planning include:

- Right-sized and future-proofed development: Ensuring adequate capacity for current and future demand.

- Modular design: Allowing for phased development and expansion.

- Incremental growth: Managing construction costs and responding to evolving needs.

- Streamlined processes: Optimizing airport layout for the efficient flow of aircraft, passengers, staff, baggage, and cargo.

- Collaborative planning: Fostering cooperation among stakeholders.

- Safety and security measures: Implementing robust standards to protect operations, passengers, and assets.

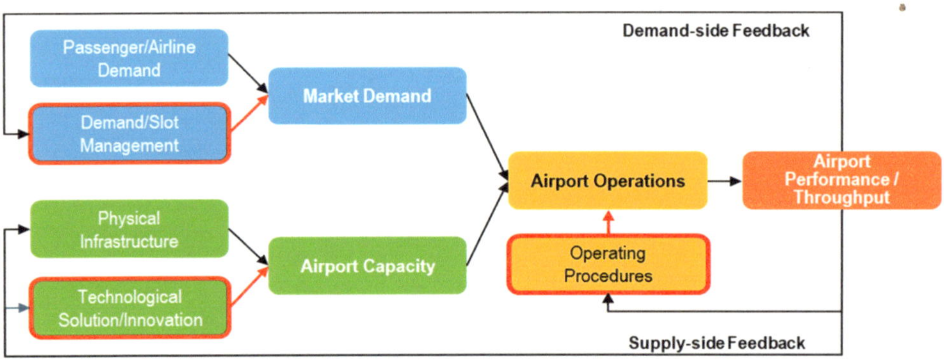

Figure 6.
Traffic demand and airport capacity alignment process. The red-highlighted activities can contribute to improving airport performance and increasing the sustainability value of airport operations, while reducing infrastructure investment requirements.

When developing optimization models, it is crucial to consider the broader airport business environment and socio-economic factors, such as ownership, population centers, and user demographics.

3.1 Research: Sydney metropolitan area airport capacity expansion case

The Sydney metropolitan area provides a compelling case study, as it encompasses all four solutions, as described in **Figure 5,** to address airport capacity challenges. As shown in **Figure** 7, the region includes the existing Kingsford Smith Airport, two noncommercial airports (Richmond Airbase and Bankstown General Aviation Airport), and the proposed Western Sydney Airport.

A combination of steady pre–COVID-19 air traffic growth, continuous urban expansion toward Western Sydney, and the current curfew restrictions at Kingsford Smith Airport – together with its annual capacity limit of 72 million passengers [6] – has underscored the need for a clear vision and strategy by the Australian government and aviation stakeholders to expand capacity in the metropolitan area.

This research aimed to develop a cost optimization model for determining the optimal approach to expanding airport capacity in metropolitan areas, accounting for demand uncertainties. The research process, detailed in **Figure 8**, involved the development of a deterministic cost optimization model using mixed-integer nonlinear programming (MINLP), incorporating six cost functions: capital cost, operational cost, delay cost, noise cost, operation readiness airport transfer (ORAT) cost, and passenger access cost, which was then extended into a stochastic optimization model. This model was validated with the Sydney case study to identify optimal solutions and investment profiles

Figure 7.
Airports and aerodromes in the Sydney metropolitan region. Geographical map displaying the locations of airports and population centers in the Sydney metropolitan area. Modified by the author using information from "Western Sydney Airport Environmental Impact Study" by Western Sydney Airport Co., 2014 [7].

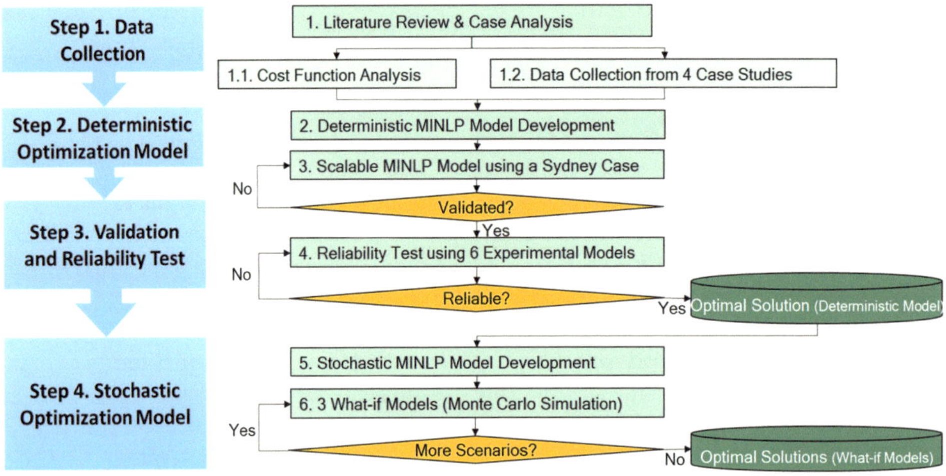

Figure 8.
The author's research process flows to develop an airport capacity expansion cost optimization model, integrating MINLP and stochastic simulation methods.

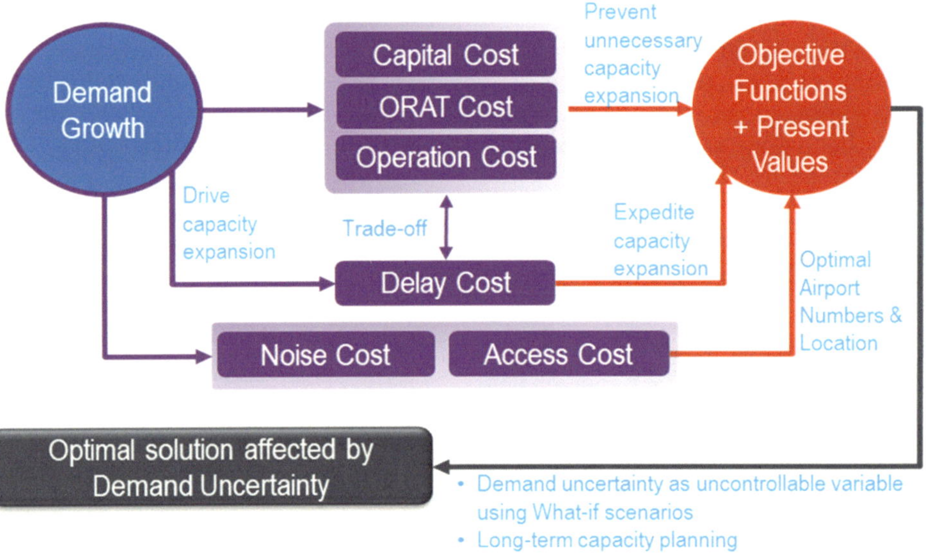

Figure 9.
Airport capacity expansion optimization modeling approach and key considerations employed in the author's research.

over a 50-year window (2011–2060). This stochastic optimization model was evaluated under three different "What-if" demand scenarios, including the impact of the COVID-19 pandemic.

As illustrated in **Figure 9**, the objective function minimizes the net present value (NPV) of total costs, encompassing the six identified cost functions. By analyzing cost–demand relationships over time and discounting future costs, the model identifies cost-minimizing solutions. Considering capital, operational, and ORAT costs helps avoid premature expansion not justified by demand. Including delay costs mitigates the risk of deferring capacity projects and the resulting congestion costs. Incorporating noise and passenger access costs supports the identification of airport locations that minimize overall costs for users and surrounding communities.

A simplified version of the objective function and mathematical formulation of the deterministic model, used to determine the decision variable for each airport's target capacity, is presented in Eq. (1). The input variables for this baseline model – including constraints and operational conditions of existing airports, passenger traffic profiles, demographic characteristics of each population center, and other model parameters – were derived from the Joint Study conducted by the Australian and NSW governments [8]. The annual traffic growth rate is assumed to remain consistent with historical air traffic trends in the Sydney metropolitan area.

$$\begin{aligned} min\, \delta_y \times \sum_i \sum_y (k_{iy} \times CC_{iy} + k_{iy} \times OPC_{iy} + k_{iy} \times DC_{iy} + k_{iy} \times NC_{iy} + k_{iy} \\ \times ORC_{iy}) + \delta_y \times \sum_i \sum_j \sum_y (k_{iy} \times AC_{ijy}) \end{aligned} \tag{1}$$

where:
CC = capital costs, costs to expand the airport's capacity.
OPC = operation costs, costs to maintain facilities and provide required services.
DC = delay costs, costs incurred from operational delay and congestion.
NC = noise costs, costs to address aircraft noise pollution.
ORC = ORAT costs, costs for commissioning new airports/new facilities.
AC = access costs, which are the costs incurred by passengers, visitors, and staff to access airports.
δ = discount coefficient.
i = *an* existing or potential airport within the metropolitan area.
y = *a* period within the planning time horizon.
J=*a* population center within the metropolitan area.
k = binary variable, indicating whether airport i provides services in period y, k = {0 or 1}.

The stochastic version of the airport capacity expansion optimization model was formulated by combining the format of the deterministic model and the multi-stage stochastic model, which can be written as Eq. (2):

$$\begin{aligned} min\, E_y \times \delta_y \times \sum_i \sum_y (k_{iy} \times CC_{iy} + k_{iy} \times OPC_{iy} + k_{iy} \times DC_{iy} + k_{iy} \times NC_{iy} + k_{iy} \\ \times ORC_{iy}) + E_y \times \delta_y \times \sum_i \sum_j \sum_y (k_{iy} \times AC_{ijy}) \end{aligned} \tag{2}$$

where:
E_y = finite random outcomes or probability variables from an event at time y, which are defined based on specific demand scenarios. In this research, three what-if scenarios were tested as outlined below:

Model	Passenger demand (in thousands, year 50)	Target capacity (in thousands, Year 50)/ operation timeline				Accumulative total costs (Year 1–Year 50)
		SYD	BWU	RCM	WSI	
Deterministic model	145,532	72,000/ Y1–Y50	Nil	Nil	73,532 / Y23–Y50	AU$25.3Bil
Stochastic Model 1	146,977	72,000/ Y1–Y50	Nil	Nil	74,977 / Y26–Y50	AU$24.2Bil
Stochastic Model 2	157,601	72,000/ Y1–Y50	3,600/ Y49–Y50	Nil	82,000 / Y30–Y50	AU$22.0Bil
Stochastic Model 3	121,866	72,000/ Y1–Y50	Nil	Nil	49,866 / Y24–Y50	AU$17.6Bil

Table 1.
Comparative analysis between the deterministic model and three stochastic models. SYD: existing Kingsford Smith Airport, BWU: Bankstown General Aviation Airport, RCM: Richmond Airbase, WSI: newly proposed Western Sydney Airport. Source: author's research paper [9].

Stochastic Model 1: Random growth of air traffic demand between 0–5% (low and upper limit).

Stochastic Model 2: Normal distribution of annual traffic growth rates based on Sydney's previous 25 years' records (no lower or upper limit).

Stochastic Model 3: Reflection of the pandemic COVID-19's impact on air traffic demand (no lower or upper limit).

Table 1 provides a comparative analysis of the outcomes from a deterministic model, and three scenario based stochastic models. The deterministic model, Stochastic Model 1, and Stochastic Model 3 suggest a dual airport solution, utilizing both Sydney Airport and Western Sydney Airport to accommodate future traffic demand for the next 50 years. In contrast, Stochastic Model 2 proposes a three-airport system, recommending the conversion of Bankstown Airport from a general aviation airport into a commercial facility to meet the rapidly increasing demand.

This variation highlights that the optimal solution for airport capacity expansion can significantly depend on the projected future traffic demand. Stochastic Model 2, which anticipates a larger target capacity for the Sydney region compared to the Deterministic model and Stochastic Model 1, requires a lower total cost over time. This model forecasts a slow growth rate in traffic demand initially, delaying the need for capacity expansion until Year 20. By postponing capital investment until the market demand justifies it, this approach effectively minimizes costs.

On the other hand, Stochastic Model 3 projects the lowest total costs due to the impact of the COVID-19 pandemic between Years 10 and 15. The significant drop in air traffic demand means that capacity expansion projects are not necessary until Year 23. The difference in air traffic demand between Stochastic Model 3 and the other models results in notable cost differences and necessary airport capacity over time.

Figures 10 and **11** illustrate significant differences in future demand and cost profiles between a deterministic model and a stochastic optimization model accounting for a pandemic-like event. While access and noise costs correlate with demand, operating costs can remain stable during demand declines due to the need to maintain infrastructure, creating inefficiencies from underutilized assets. The

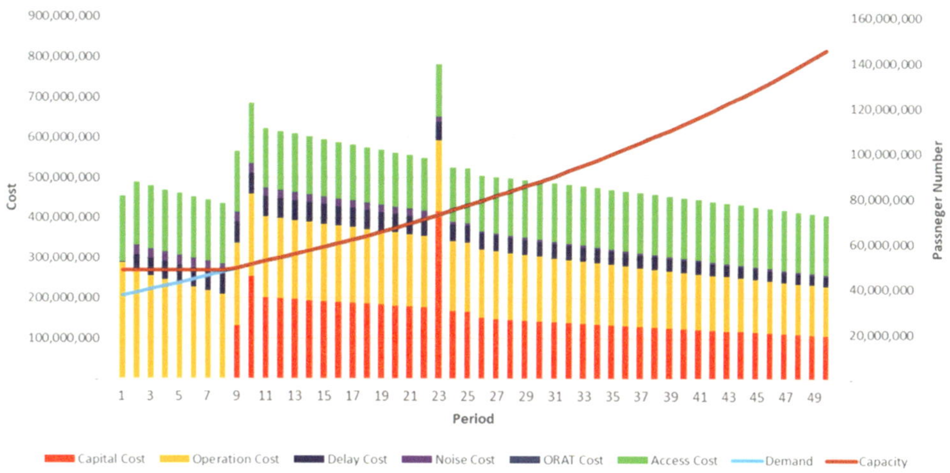

Figure 10.
Demand–capacity–cost comparison: Sydney deterministic model. Source: author's research paper [9].

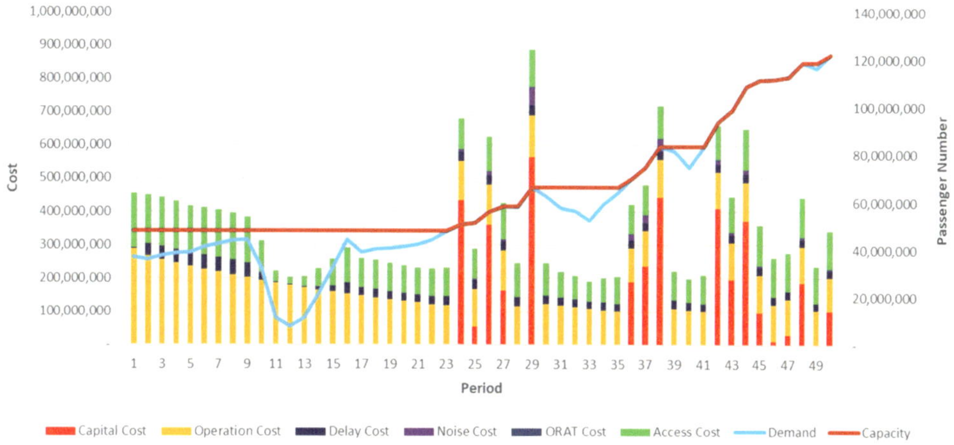

Figure 11.
Demand–capacity–cost comparison: Sydney Stochastic Model 3 (pandemic event). Source: author's research paper [9].

models suggest that the optimal timing for Western Sydney Airport varies with market conditions but remains a strong long-term solution. This demonstrates the importance of stochastic optimization to account for uncertainties and optimize long-term infrastructure investments.

This study contributes to the understanding of how airports in metropolitan areas interact with urban economies and social dynamics, and it introduces a decision model for sustainable airport capacity expansion. By integrating airport connectivity, delay, and noise considerations, the model enables a more strategic approach to advancing airport stakeholders' sustainability goals.

The cost optimization decision model can also be adapted to address additional real-world challenges by modifying the objective function and incorporating further constraints, particularly for small- and medium-sized airports. For example, it could be extended to optimize social benefits, passenger throughput, or commercial revenues. Future iterations of the model might also account for uncertainty factors such as competition from high-speed rail, financial crises, political developments, airline competition, technological innovations, and demographic shifts.

4. Digital transformation for sustainable airport development and operations

The International Air Transport Association (IATA) projects that global passenger volumes will double by 2041, placing significant strain on existing airport infrastructure and processes [5]. Given the complexities and long lead times associated with physical capacity expansion, the aviation industry must increasingly rely on automation, digitalization, and process optimization to manage this growth sustainably. Despite the adoption of self-service technologies such as online check-in, automated bag drops, and biometric gates, the passenger journey remains fragmented and often involves manual document checks. IATA's 2024 Global Passenger Survey revealed that this reliance on outdated processes contributes to persistent congestion and long queues at key checkpoints, with a reported 37% of travelers citing queue times as a major frustration [10].

The technological groundwork for a more seamless experience is already in place. Passenger willingness to embrace digital, mobile-first solutions is high, with 89% open to sharing immigration data to expedite processes and 84% expressing satisfaction with biometric identification [10]. As Alan Murray Hayden, IATA's Director for Airlines, Airports, and Security Solutions, noted, "Passengers have made it clear they want to spend less time queuing and moving through the airport faster. And they are increasingly willing to use biometric data to complete more pre-departure tasks off airport to achieve this." However, while many airports have implemented COVID-19 mitigation measures and are considering introducing more robust, innovative technology solutions, a large number still lack long-term resilience plans, and most continue to face challenges in technological readiness for future disruptions [11].

Digital aviation, powered by data analytics and AI, has strong potential to transform the passenger experience via digital identities and biometric processing. The European Union's forthcoming Digital Travel Credential (DTC), expected around 2030, will allow passengers to use their faces as primary ID, streamlining identification and security. Advances in pre-travel admissibility – completing immigration and visa requirements online – are poised to reduce bottlenecks and the risk of denied boarding. In Southeast Asia, hubs such as Singapore and Bangkok are piloting biometric e-gates and digital visa platforms, though regional harmonization remains a challenge.

The COVID-19 pandemic further accelerated the adoption of digital technologies for operational resilience and public health. Airports implemented a range of innovative solutions, including:

- IoT and web-service platforms optimize HVAC systems and monitor volatile organic compounds (VOCs), thereby improving ventilation and reducing the risk of cross-contamination.

- Advanced camera systems for passenger monitoring, enabling queue reservation and virtual queuing.
- AI-driven video infrastructure at primary monitoring points supports the prevention of communicable disease spread.
- Detailed passenger flow analysis to optimize infrastructure capacity and minimize cross-contamination between low- and high-risk passenger groups.
- Replacement of touchpoints with facial recognition and contactless payment systems.
- Digital way-finding provides real-time information on directions and queue times.
- Mobile-based digital health certificates and travel credentials for verification.

Despite opportunities, the industry faces challenges in achieving full digital transformation and data integration (**Table 2**):

Opportunities	Challenges
End-to-end passenger journey optimization: AI-enabled biometrics, touchless processing, and predictive analytics to improve efficiency. Integrated mobility across ground, air, and sea transport.	**Data governance and privacy:** Concerns over data ownership, security, and regulatory compliance, particularly regarding passenger biometrics.
Airside and operational efficiency: AI-driven decision support for aircraft turnaround, stand allocation, and baggage handling. Real-time data integration for enhanced safety and reduced delays.	**Legacy systems and infrastructure:** Outdated IT systems that limit interoperability and hinder the adoption of new digital platforms.
Resources and sustainability gains: Smart energy management, predictive maintenance, and optimized scheduling to reduce environmental footprint and costs.	**Scalability and standardization:** AI pilots often remain small-scale due to a lack of industry-wide standards, which makes replication difficult across different airports or various parts within the airport.
Collaborative data platforms: Shared digital ecosystems (e.g., Airport Operations Control Centers) enable transparent decision-making. Cloud-based data integration for scalable AI deployment.	**Fragmented stakeholder ecosystem:** Data silos and differing standards exist among airlines, border agencies, and ground handlers.
New revenue and service models: Personalized services, dynamic retail offerings, and data-driven insights to create nonaeronautical revenue. AI to support the evolution into integrated urban platforms.	**Change management and culture:** Stakeholder resistance arises due to concerns about disruption, job impacts, and an unclear return on investment (ROI).

Table 2.
Opportunities and challenges of digital transformation for sustainable airport development and operation.

4.1 Global movement: ICAO GISS 2024/2025 – the role of AI and technologies

The International Civil Aviation Organization (ICAO) has placed aviation sustainability and digital transformation at the forefront of its agenda. The ICAO Global Implementation Support Symposium (GISS) has been a key forum for these discussions. While the 2024 GISS established a strong foundation for global partnerships and sustainable financing [12], the 2025 symposium [13] specifically highlighted the pivotal role of AI in driving technology-led sustainability. This marked a significant shift, underscoring AI's practical applications in operations, training, and multi-stakeholder coordination. The symposium's agenda focused on how advanced technologies are reshaping airspace and airport management, enhancing safety, and advancing environmental stewardship, all while supporting ICAO's "No Country Left Behind" initiative.

AI has emerged as a practical tool for streamlining airport operations and supporting digital transformation on a large scale. A prime example is the use of AI in terminal energy management systems.

4.2 Case study: AI in terminal energy management systems

AI-driven energy management enables airports to improve efficiency, reduce costs, and enhance sustainability. By leveraging predictive modeling and smart control systems, airports can achieve proactive and responsive energy use. The core concept involves processing real-time data from sensors, along with information on weather forecasts, energy demand, and passenger concentration through specific AI algorithms. The system then constantly adjusts building elements to maintain an optimal temperature and comfort level while minimizing energy consumption. AI provides a powerful toolkit for optimizing energy use, offering several key benefits:

- Predictive energy modeling: Machine learning analyzes historical data to forecast energy demand, enabling proactive load management and reducing costly peak demand charges.

- Intelligent building management systems (i-BEMS): AI-driven solutions integrate HVAC, lighting, and water systems, dynamically adjusting operations based on real-time data to reduce waste.

- Adaptive control of renewable energy assets: AI optimizes the integration of solar, wind, or waste-to-energy sources, balancing energy flows for cost-effectiveness and sustainability.

- Anomaly detection and demand alerts: AI platforms monitor energy systems to detect inefficiencies or leaks early, prompting timely corrective actions.

Recent projects demonstrate the practical application of this technology:

- Cologne/Bonn Airport implemented an AI pilot to optimize its whole-site energy supply, enhancing real-time source selection and reducing on-site emissions [14].

- Jewel at Changi Airport deployed an AI-driven energy intelligence platform to address overcooling issues, successfully optimizing energy usage while maintaining occupant comfort. [15].

- Istanbul Airport used AI-driven analysis to identify solar microgrids and AI-managed energy systems as top sustainability investments, based on a multicriteria decision model that considered environmental, financial, and operational returns [16].

5. Conclusions

To accommodate projected growth in air traffic, all aviation stakeholders must work collaboratively. Sustainable airport capacity expansion is critical in the post-pandemic era, as it supports both the aviation industry and broader urban development by addressing evolving travel demand and technological advancements. As a key driver of economic growth, aviation facilitates trade, tourism, and connectivity. Therefore, adaptable airport infrastructure is essential for handling fluctuating traffic, integrating new technologies, and minimizing environmental impact, thereby fostering both industry prosperity and urban progress.

In a new era of growth, the industry's trajectory will be increasingly shaped by challenges such as economic uncertainty, geopolitical tensions, and capacity constraints. While advanced markets may face tempered growth due to supply chain bottlenecks and capacity shortages, emerging markets are likely to see continued expansion driven by higher infrastructure investment and a rising middle class. To navigate this landscape, airports must prioritize financial viability, infrastructure investment, operational efficiency, and sustainability.

The following are key strategies for achieving sustainable and smart airport capacity expansion:

- Advanced planning and modeling: Employ machine learning and stochastic modeling to predict traffic patterns under various scenarios. Use data analytics to inform and adjust capacity planning.

- Optimized infrastructure: Implement modular construction and adaptable designs (e.g., swing gates) to allow for flexible expansion. Utilize intelligent air traffic management systems to optimize flight paths and reduce delays. Design building envelopes to be efficient and safeguard space for future technologies.

- Sustainable operations: Invest in renewable energy, implement water conservation and waste management programs, and promote sustainable transportation options for passengers and employees.

- Smart airport management: Utilize biometric technology for streamlined security, implement digital way-finding, and employ predictive maintenance to minimize downtime.

- Stakeholder engagement: Foster collaboration among airlines, air navigation service providers, and other partners to ensure a coordinated approach and to share valuable data. Engage with local communities to address concerns related to noise and environmental impact.

In conclusion, efficient use and integration of data are paramount for understanding passenger behavior and operational patterns. Airports must embrace data-driven insights in their planning, design, and operations. By acting as catalysts for the entire air transport system and working in close coordination with all operational stakeholders, airports can play a critical role in successful and sustainable capacity expansion and management.

Acknowledgments

I would like to extend my sincere gratitude to my colleagues at Bechtel, whose expertise, dedication, and collaboration have been invaluable throughout the development of the Western Sydney International Airport. Their commitment to excellence and teamwork has been a constant source of support and inspiration.

I am equally grateful to the WSA customer team, whose vision, guidance, and partnership have been instrumental in achieving the project's successful outcomes. Working alongside such talented and committed professionals has been a privilege, and this book would not have been possible without their contributions.

The authors acknowledge the use of ChatGPT for language editing in the chapter.

Conflict of interest

The authors declare no conflict of interest.

Author details

Woojin Choi
College of Aviation & Management, Korea Aerospace University, Seoul, South Korea

*Address all correspondence to: iwoojin@kac.ac.kr

References

[1] ACI World, Exclusive preview of the 2024 airport rankings [Internet]. 2025. Available from: https://aci.aero/2025/04/14/the-busiest-airports-in-the-world-defy-global-uncertainty-and-hold-top-rankings [Accessed: 2025-August-24]

[2] ACI World, ACI World Airport Traffic Forecasts 2024–2053 [Internet]. 2025. Available from: https://store.aci.aero/wp-content/uploads/2025/02/World-Airport-Traffic-Forecasts-2024-2053-Executive-Summary.pdf [Accessed: 2025-August-18]

[3] WSA Co Limited, Sustainability Plan - Design and Construction. 2022. Available from: https://westernsydneyinternational.widen.net/s/wwxf2vhc6g/wsa-sustainability-plan-summary-2022 [Accessed: 2025-June-15]

[4] ACI World Insight, Case study: Checkpoint 90 at Schiphol airport. 2025. Available from: https://blog.aci.aero/environment-and-sustainability/waste-not-want-not-uncovering-the-possibilities-of-urban-mining-in-aviation/ [Accessed: 2025-August-2]

[5] Hub IATAK, Transforming the travel experience: What are the solutions for future congested airports? [Internet]. 2023. Available from: https://www.iata.org/en/publications/newsletters/iata-knowledge-hub/ [Accessed: 2025-July-17]

[6] Sydney Airport Co Limited, Sydney Airport Master Plan 2045, 2025. Available from: https://www.masterplan2045.com.au/ [Accessed: 2025-August-01]

[7] Australian Government, Environmental Impact Assessment – Western Sydney Airport [Internet]. 2015. Available from: https://www.westernsydneyairport.gov.au/media-resources/resources/environmental-assessment [Accessed: 2025-July-23]

[8] Australian and NSW Government (2012). Joint study on aviation capacity in the Sydney region (Technical Paper - Volume 3 and Volume 5). Available from: https://www.westernsydneyairport.gov.au/sydney_av_cap [Accessed: 2025-June-15]

[9] Choi W, Truong, D. Cost optimization modeling for airport capacity expansion problems in metropolitan areas. Journal of Airport Management;**19**(4):373–392. Available from: https://hstalks.com/article/9986/cost-optimisation-modelling-for-airport-capacity-e/?business DOI: 10.69554/MVPL7588

[10] IATA Global Passenger Survey (GPS) [Internet]. 2024. Available from: https://www.iata.org/en/publications/manuals/global-passenger-survey/ [Accessed: 2025-July-05]

[11] Ammoury M, Salman B. Advancing sustainability and resilience of airports through deployment of new technologies in the aftermath of the COVID-19 pandemic. Multidisciplinary Journal of Civil Engineering. 2024;**2**(1). DOI: 10.1061/AOMJAH.AOENG-0027

[12] International Civil Aviation Organization, ICAO Global Implementation Support Symposium 2024: Optimizing Aviation Capacity for a Sustainable Future [Internet]. 2024. Available from: https://www.icao.int/news/icao-global-implementation-

support-symposium-2024-optimizing-aviation-capacity-sustainable [Accessed: 2025-August-19]

[13] International Civil Aviation Organization, ICAO Global Implementation Support Symposium 2025: The Next Flight: Innovating for a Sustainable Tomorrow [Internet]. 2025. Available from: https://www.icao.tv/global-implementation-support-symposium/season:4 [Accessed: 2025-July-28]

[14] Cologne Bonn Airport, Pilot project for efficient digital energy management [Internet]. 2024. Available from: https://www.cologne-bonn-airport.com/en/company/newsroom/press-releases/detail/pilot-project-for-efficient-digital-energy-management.html [Accessed: 2025-July-13]

[15] Resync, Jewel at Changi Airport Partners with Resync to Elevate Energy Efficiency and Passenger Comfort [Internet]. 2025. Available from: https://www.resynctech.com/jewel-at-changi-airport-partners-with-resync-to-elevate-energy-efficiency-and-passenger-comfort/ [Accessed: 2025-August-09]

[16] Mizrak F, Sahin DR. Investment strategies for renewable energy technologies and harvesting systems in airport operations using spherical fuzzy MCDM models. Scientific Reports. 2025;**15**(1):24075. DOI: 10.1038/s41598-025-08480-7